Carceral Space, Prisoners and Animals

T0203734

Carceral Space, Prisoners and Animals explores resonances across human and nonhuman carceral geographies. The work proposes an analysis of the carceral from a broader vantage point than has yet been done, developing a 'trans-species carceral geography' that includes spaces of nonhuman captivity, confinement, and enclosure alongside that of the human. The linkages across prisoner and animal carcerality that are placed into conversation draw from a number of institutional domains, based on their form, operation, and effect. These include: the prison death row/ execution chamber and the animal slaughterhouse; sites of laboratory testing of pharmaceutical and other products on incarcerated humans and captive animals; sites of exploited prisoner and animal labor; and the prison solitary confinement cell and the zoo cage. The relationships to which I draw attention across these sites are at once structural, operational, technological, legal, and experiential / embodied. The forms of violence that span species boundaries at these sites are all a part of ordinary, everyday, industrialized violence in the United States and elsewhere, and thus this 'carceral comparison' amongst them is appropriate and timely.

Karen M. Morin is Associate Provost and Professor of Geography at Bucknell University, in Lewisburg, Pennsylvania, U.S. Her interests span the history of geographical thought in North America, 19th-century travel writing, post-colonial geographies, carceral geography, and critical animal studies. She is author of *Frontiers of Femininity: A New Historical Geography of the Nineteenth-Century American West* (2008) and *Civic Discipline: Geography in America, 1860–1890* (2011). She is co-editor of *Women, Religion, and Space: Global Perspectives on Gender and Faith* (2007) and *Historical Geographies of Prisons: Unlocking the Usable Carceral Past* (2015).

Routledge Human-Animal Studies Series

Series edited by Henry Buller
Professor of Geography, University of Exeter, UK

The new *Routledge Human–Animal Studies Series* offers a much-needed forum for original, innovative and cutting-edge research and analysis to explore human–animal relations across the social sciences and humanities. Titles within the series are empirically and/or theoretically informed and explore a range of dynamic, captivating and highly relevant topics, drawing across the humanities and social sciences in an avowedly interdisciplinary perspective. This series will encourage new theoretical perspectives and highlight ground-breaking research that reflects the dynamism and vibrancy of current animal studies. The series is aimed at upper-level undergraduates, researchers and research students as well as academics and policy-makers across a wide range of social science and humanities disciplines.

Carceral Space, Prisoners and Animals

Karen M. Morin

Routledge
Taylor & Francis Group

LONDON AND NEW YORK

First published 2018 by Routledge

2 Park Square, Milton Park, Abingdon, Oxfordshire OX14 4RN

52 Vanderbilt Avenue, New York, NY 10017

Routledge is an imprint of the Taylor & Francis Group, an informa business

First issued in paperback 2020

British Library Cataloguing in Publication Data
A catalogue record for this book is available from the British Library

Library of Congress Cataloging in Publication Data
A catalog record for this book has been requested

ISBN: 978-1-138-63987-4 (hbk)
ISBN: 978-0-367-35999-7 (pbk)

Typeset in Bembo
by Taylor & Francis Books

Contents

Figures

Prologue

As a scholar of American West history and geography I have always been attuned – or so I thought – to the many actors and agents who encountered one another on the western U.S. frontier in the 19th century – among them Native peoples, escaped slaves, White settlers, farmers and miners, tourists and travelers, Army recruits, and colonial bureaucrats. One book that informed my early research and teaching on the interactions of these groups was Robert Dykstra's *The Cattle Towns* (1968). The book examines the development of small western towns in the 1870s and 1880s, towns established along the route of the free-range cattle trade by small groups of local entrepreneurs. Cattle towns such as Dodge City, Kansas lay at the junction of livestock trails and the railroad, and they provided facilities to receive and sell herds driven up from the South, especially Texas, and a transportation hub connected to ranches and meatpackers in Chicago and beyond. Years ago I would have critiqued Dykstra's book in my analysis of gender norms and colonial relations in the American West (Morin 2008). Today, I wonder more about the cattle, and the vast industrial infrastructure that developed to exploit their labor and make their bodies into meat. Dykstra's themes of commercial development (cattle meat trade), cultural change (saloons, gambling, prostitution), in-migrations of transient worker populations (cowboys), and the rapid expansion of town police forces strikes me now as disconcertingly similar to other small, rural archipelagoes developed on the 'trade' in other lives exploited for profit – U.S. prison towns.

In the prison town of Susanville, California, for example, capital investment by local officials and entrepreneurs, expansion of the prison infrastructure on the built environment, and the increasingly familiar, everyday impact of the prison on social relations and cultural norms is obvious (Fraser 2000). Local shops, bars, fuel stations, and workers who quickly move through Susanville on their career advancement trajectory all shape this prototypical prison town that is like so many others across the U.S. landscape (Eason 2010; Che 2005; Bonds 2009). In the United States the prison boom over the past four or five decades brought a 'prison archipelago' of correctional facilities to innumerable, often rural, settlements like Susanville as part of their economic development strategies. And just as the violence wrought by the prison town is mostly hidden behind prison walls, most 'frontier violence' of the cattle towns was enacted not through the cattlemen's alcohol consumption and gun violence as Hollywood may wish us to believe, but rather via colonial wars with and land dispossession of Native peoples (Wishart 2004: 162, 387) as well as the 'invisible' violence effected on the cattle themselves.

Studying in detail the respective historical-geographical developments of the cattle town and the prison town will make a fitting *next* project for me. I mention the juxtaposition of these towns and their overlapping social, political, and economic themes by way of offering a starting point and segue into the 'carceral logics' underlying this book. That is to say, *Carceral Space, Prisoners and Animals* places into parallel conversation the connected and entangled spatial, structural, operational, and embodied carceral practices and processes of a number of industrial sites and institutions in the present-day U.S. While I touch on the respective historical-geographical genealogies of the sites I study here – the prison, slaughterhouse, research lab, zoo, and farm – my project is to uncover the epistemic violence that pervades contemporary industrial America yet is normalized and 'neutralized' in countless ways in everyday life. This book is more conceptual than it is theoretical, in that I engage some of the most compelling of the current themes, arguments, and activist scholarship of carceral geography and critical animal studies. My intention is to broker between these ideas and literatures, bring them into conversation, and through this process highlight the ways that violence within U.S. industrial regimes are enmeshed and maintained together.

As such, this was a dark and difficult book to live with. The embodied experiences of countless human and nonhuman lives subjected to desperate, ghastly conditions were brought into my radar and became personal to me in the past few years in ways that I had clearly disavowed in the past. I felt many times that I was experiencing the inescapability of a sort of clockwork-orange-like rehabilitation. If, as Donna Haraway (2008: 78) wryly claimed, "everyone may know, but there is not nearly enough indigestion," I will say that in my case, the level of indigestion has been made sufficiently high through the writing of this book.

Over the past decade and a half I have been involved with a local prisoner rights group, the Lewisburg Prison Project, and I have to go back to my graduate school days at the University of Nebraska–Lincoln to fully appreciate where the inspiration for this work originated. That is, with Frances W. Kaye, an English professor I greatly admired and worked for as editorial assistant at the *Great Plains Quarterly*, who, for decades, protested against the death penalty in front of the governor's mansion in Lincoln, every Wednesday at noon, usually by herself. (Fortunately for Nebraska, Fran is still around, considering the current governor's somersaults to keep the death penalty alive after a legislative ban was enacted last year.) Meanwhile, my work with colleagues at the Prison Project and letters from thousands of prisoners incarcerated in the Middle District of Pennsylvania who we attempt to serve, provide a constant ringing of the alarm bell to the horrific nature of contemporary "corrections." I thank in particular Dave Sprout and Elayne Sobel for tracking down some valuable information I draw upon here. It wasn't until a visit to New York's Bronx Zoo celebrating my daughter's birthday a few years ago, though, that the ideas for this book began to take shape. Watching the crowds of tourists, most but not all of them children, jump up and down and pound the glass wall of one particular Amur tiger's cage struck me in ways that visits to zoos had not in the past; while the beautiful creature paced anxiously around and around a big rock, the notion of a trans-species carceral geography really clicked in.

I was immensely lucky to participate in the Race and Animals Institute at Wesleyan University in the summer of 2016, led by Lori Gruen, Claire Jean Kim, and Timothy Pachirat. Timothy in particular sparked insights for this book, and I also want to thank him for permission to reproduce one of the maps that appeared in his extraordinary *Every Twelve Seconds* (2011). Actually, participation at the Institute gave a coherence to this project that it simply would not have otherwise had, although of course errors of interpretation remain my own. Guest lectures and presentations at the Institute were given by Jared Sexton, Maria Elena Garcia, Colin Dayan, and fellow participants Nekeisha Alayna Alexis, Brigitte Fielder, Che Gossett, Alexandra Isfahani-Hammond, Kelly Struthers Montford, Rachel Mundy, Fiona Probyn-Rapsey, Calvin John Smiley, and Dinesh Wadiwel. Truly, I had no idea what I didn't know until participating at the Institute, and as it turns out, I didn't know most things relevant to this project. This is a long way of saying that the Institute deeply challenged me and I want to again thank Lori and all the leaders and partici-pants for so graciously nudging me without judging me. Harriet Washington's visit to my university a few months ago also inspired me, at just the right moment in my writing. I also presented portions of this work as it developed at conference sessions over the past couple of years, challenged in countless ways by my fellow presenters and audiences – at the Critical Interventions in a Multi-Species World panel at the Association of American Geographers meeting in Chicago (2015); the Neue Kulturgeographie XIII Conference in Graz, Austria (2016); the 7th Nordic Geographers' Meeting in Stockholm (2017); and the Minding Animals International Conference in Mexico City (2018). I offer a special thanks to my sometimes collaborator Dominique Moran, who is always one step ahead in her enlightening and creative thinking.

The individual to whom I owe the greatest debt for assistance with this project is the formidable Kathryn Gillespie, who read the entire manuscript, more than once. I will never be able to thank Katie enough for the incredibly generous and insightful feedback she provided, for preventing me from making countless uninformed mistakes, and for memorably sharing lunch at Crakovia one day, which featured beets served six ways. She and Rosemary-Claire Collard kindly published my chapter "Wildspace" (a version of which appears here as Chapter 5), in their terrific collection *Critical Animal Geographies* (2015). I am also most grateful for Khalil Saucier's astute reading of the entire manuscript, offering feedback that was fantastically helpful at key junctures. The vast intellectual network of geographers and animal studies scholars who further influenced this work will be evident in the relentless citations throughout.

The long genesis of this project, and the 'distilled' journal article version of it published earlier (Morin 2017), means that I owe additional thanks to many individuals who helped me sharpen my thinking early on, especially the *Antipode* editors and staff. I enjoyed several fruitful conversations with my Bucknell colleague Gary Steiner, and he generously read drafts of my work and shared valuable reading lists. His colleagues in Bucknell's Philosophy Department invited me to share the work in a Lunch Chat, which, among other things, guided me to thinking again about the Ota Benga story and the 'human zoo' display at the Bronx Zoo Monkey House ca. 1906 (Chapter 3). Carol White's provocative work in recuperating human animality (e.g. White 2017) will be of interest to anyone reading this book. Thanks once more to Bucknell's virtuoso videographer Brianna Derr; the talented graphic designer Josie Fertig; and digital specialist Debra Cook-Balducci, who has provided image production help with innumerable of my projects over the years and once again came through generously, thoughtfully, and expertly. The institutional support I have continually received from Bucknell University – in time and other resources – I am profoundly grateful for. And Faye Leerink, Ruth Anderson, Kelly Cracknell, and the entire production team at Routledge have been tremendous allies in seeing this book to print.

Finally, special thanks to my friends and family who patiently heard more about this project than maybe they cared to – Mark Mattern, Katherine Kratz, Maria Antonaccio, Annie Randall, my illustrious book club friends Libby, Stacy, Susan, and Toni – thanks to you all for sharing in this dark matter and occasionally checking in on its progress. My indefatigable sister Ann Baker challenged my thinking in more ways than she knows. My children Nina Theresa and Nicholas Hunter, I can't say how much love for you there is in this work. Our family's best friends, Bobby Lopez, Mia, Mel, Templeton, and Rosie before them, were always with me in this project too, in one way or another. This book is dedicated to Daniel Olivetti, *still* my favorite person, who not only kindly read the entire manuscript and provided great advice from the point of view of an art gallery director – and who doesn't need that – but whose cordially-acquired taste for polenta and zucchini hasn't gone without notice.

1 Introduction

Carceral space, prisoners and animals

Introduction

Standing at the top of the most iconic place on my university campus – the quad – one overlooks a beautiful pastoral landscape that encompasses the lower part of campus to the far extent of the ridge-and-valley Appalachian Mountains in the distance. This otherwise picturesque view of Pennsylvania's rolling hills and steepled architecture is always spoiled for me, though, by two sites of enclosure poking out of the landscape: the upper portions of the "monkey cages" facility that are the hallmark of the university's animal behavior program, and the bell tower of the federal penitentiary, USP-Lewisburg, on the edge of town. The monkey cages facility currently houses an impressive array of primates for an undergraduate liberal arts school, including baboons, lion tail macaques, squirrel and capuchin monkeys, all of whom live out their lives in what appear to be small, mostly concrete lab cages. Few in the community would notice or care that the lived experience of these animals is not unlike that of the 1,000+ prisoners at the penitentiary, a site that has become notorious in recent years for its insidious Special Management Unit (SMU) program that features 23- or 24-hour a day double-celled lockdown of federal prisoners from around the country, brought here for years of "readjustment" (Morin 2013).

The juxtaposition of the Lewisburg penitentiary and the animal research labs at my university offers a useful leitmotif for the subject of this book. The nexus of mass incarceration in U.S. prisons and mass exploitation of nonhuman animals today presents a "critical moment in history" (Thomas and Shields 2012: 4). These processes are connected, and the purpose of this book is to develop a framework to position and interpret some key points of connection across these human and nonhuman 'carceral spaces'. Although incarceration has conventionally come to refer to the legal confinement of prisoners under the jurisdiction of the state, 'the carceral' has also come to be understood as embracing the myriad ways in which persons could be confined by other means, such as in spaces of detention for immigrants and refugees, as well as those 'trans-carceral' spaces touched by the prison and security state apparatus outside of the formally carceral, that spill over into everyday life in myriad ways (after Foucault 1977;

Peck 2003; Moran et al. 2013; Loyd et al. 2012; Shabazz 2015a). In this work I propose an analysis of the carceral from a broader vantage point than has yet been done, developing a 'trans-species carceral geography' that includes spaces of nonhuman captivity, confinement, and enclosure alongside that of the human. As I discuss in the next section, Moran et al. (2017) advance a useful taxonomy, if you will, to define carceral space as encompassing a set of 'carceral conditions' that bear the nature and quality of carcerality, and in so doing help us move beyond Foucault's (1977: 298) notion of carcerality as encompassing the social in its entirety.

The linkages across prisoner and animal carcerality that I place into parallel conversation draw from a number of institutional domains, based on their form, operation, and effect. These include: the prison death row/execution chamber and the animal slaughterhouse (Chapter 2); sites of laboratory testing of pharmaceutical and other products on incarcerated humans and captive animals (Chapter 3); sites of exploited prisoner and animal labor (Chapter 4); and the prison solitary confinement cell and the zoo cage (Chapter 5). The relationships to which I draw attention across these sites are at once structural, operational, technological, legal, and experiential/embodied. The forms of violence that span species boundaries at these sites are all a part of ordinary, everyday, industrialized violence in the United States and elsewhere, and thus this 'carceral comparison' amongst them is appropriate and timely.

Acampora (2006), citing the work of philosopher Heini Hediger, first drew my attention to the potential of a cross-species 'carceral milieux'. In his study of zoos Acampora (2006: 109) briefly noted the similarities among calls for the reform of zoos and the reform of prisons – that is, progressives' calls for improvement of prisons resonated with the professionalizing practice of zoo management. In both cases, Acampora argued, the reformer "wants the inmates to feel as comfortable, as snug, and as much at home as possible" – with such comfort perhaps belying nothing more than increasingly sophisticated means of regulating and disciplining captive bodies. Temple Grandin (2012), for example, a spokesperson popular with the global meat industry, employs a 'somatic sensibility' in designing putatively 'humane' slaughterhouses that reduce the stress, anxiety, and suffering of the animals passing through them. In instances such as this we see an argument to reform and improve the conditions of captivity and death, which stands in stark contrast to arguments for abolishing institutions like the slaughterhouse altogether (a theme I return to in the Afterword). Acampora's insights about the cross-species carceral milieux and their respective politics and policies challenged me to think about other sites and institutions of human and nonhuman carcerality that could offer further useful insights into regimes of captivity, structures of oppressions and inequalities, and epistemologies of violence.

The sites and institutions of human and nonhuman incarceration I discuss are embedded within the Prison Industrial Complex (PIC), the Agricultural Industrial Complex (AIC), and the Medical Industrial Complex (along with a key site of the 'Entertainment Industrial Complex', the zoo). Thus, I bring together and into conversation the insights emerging from carceral geography and criminology with those of critical animal studies. I synthesize works that address human and nonhuman captivity and disposability within the PIC, MIC, and AIC – the ways that humans and nonhumans can be made disposable and killable in the prison and slaughterhouse; can be exploited for entertainment or as experimental research material; and whose bodies and labor can be made into property and commodity. My theoretical and conceptual focus is on the geographies of these sites (locations, design, and layout); the highly regulated technologies and movements within them; the emotional and psychological strain enacted via daily operations; the legal contexts within which these industries are (or are not) regulated; and the ways in which 'animalization' of certain bodies works to create the conditions for their exploitation and disposability. These are not particularly new ideas, but I offer a novel synthesis and application of them. Cross-pollinating these themes and literatures and brokering between them offers an opportunity to reflect not only on the ways in which industrial violence against humans and nonhuman animals is naturalized and made possible, but also the ways that these regimes of violence are maintained *together* – they are enmeshed and entangled in similar processes, co-constituted and co-articulating in their basic carceral logics.

The developmental trajectories and overlaps of these industrial complexes followed different historical-geographical paths towards their present iterations. With a few exceptions, my project is not in specifically comparing these developmental trajectories – governmental regulations or deregulations of industrial processes; legal maneuverings that served to protect, enrich, and incentivize certain practices; architectural or technological advances; or so on. Nonetheless a number of geographical and historical developments cannot go without notice – such as the rise of the prison industry that in many ways stepped in to fill the economic void created in many communities by deregulation of factory-scale farming. More to my project though are what these industries collectively produced over the past few decades: unprecedented numbers of confined bodies subjected to unprecedented levels of violence within the industrial U.S. today.

Key to my thesis throughout is that the distinctions between 'the human' and 'the animal' themselves are made *through* encounters with carceral spaces. Which humans and nonhumans have the force of legal, political, cultural, or other protections due to their special 'human' qualities, and which fall outside of those protections as 'animal'? The process of 'animalization' in particular subjugates both certain humans and certain nonhumans into hierarchies of worthiness and value. Fundamental to how and why certain prisoners and certain animals can be exploited, objectified, or made killable within the prison, the farm, the research lab, and the zoo are the social constructions of the human–nonhuman divide – the 'carceral logics' and social meanings that attach to various bodies and populations. The hierarchies that these distinctions perpetuate are based on a number of social markers, perhaps most importantly, racial ones. Racial difference is foundational, for example, to much of the 'criminal as animal' rhetoric, particularly via animalistic representations of Black and other minoritized men

(Kim 2016; Deckha 2010; Alexander 2012; Cacho 2014; Wacquant 2001). Meanwhile certain animals such as pets can be anthropomorphized and 'humanized' while others – vermin, pests, livestock – remain 'animalized'. Many processes are in play that either amplify the status of certain humans and nonhumans, or reduce the status of others. These have different and important implications – not least of which are the processes that govern how certain lives can be made disposable and killable because they lack ostensibly human qualities. 'The human', though, is itself a highly contested category, from which many human lives have been and continue to be excluded (Wynter 2003). And indeed, perhaps in Western societies at least, it has only been the White, Western, bourgeois man who has ever occupied or been imputed the place of 'complete' human (Ko 2016; Jackson 2015; see below and Chapter 3 for further discussion).

In what follows I do not argue that the carceral oppressions experienced by various vulnerable populations are exactly the same, but neither do they need to *be* the same in order to 'think them together', to observe, juxtapose, associate, and ultimately challenge the disciplinary regimes and structures of violence that comprise these often taken-for-granted industrial sites. Quite obviously these sites had and have distinct historical and geographical developments and contexts and are characterized by unique forms of oppression that impact lives very differently. To highlight their similar structures and mechanisms is not to say that these abuses are the same – "different atrocities deserve their own languages" (Haraway 2008: 336N23). I acknowledge their important differences, for example the sheer scale of the number of bodies exploited, abused, and killed within these spaces is worth considering. Many would argue that the thousands of prisoners on death row or executed via capital punishment does not compare to the billions of nonhuman animals killed in the U.S. each year for food or other commodities (Chapter 2). The issue of 'informed consent' across human and nonhuman groups to labor or serve as research subjects for the profits of others is a highly contested issue (Chapters 3 and 4). And clearly the purpose or 'intent' of the prison, the farm, the slaughterhouse, the zoo, and the research lab are different – they are marked by different purposes and end 'products'. Most spaces of animal confinement and captivity were not invented or designed to be places of punishment. And (at least arguably) the prison was not invented or designed to primarily provide a cheap source of labor for the manufacture of textiles (Chapter 4) or to assemble a captive population on which to test infectious diseases (Chapter 3).

Nonetheless, I would argue that much can be gained by placing these sites in conversation and highlighting their common denominators, the ways that their processes are entangled and reinforce one another. I bring together insights from carceral geography and critical animal studies to expose how the politics and regimes of domination work throughout some of the main industries in the U.S. today, reaching new levels of "industrialized efficiency" (after Gilmore in Loyd 2012: 42). It is important to be able to recognize the carceral logics, respective taxonomies of power that work through hierarchies of 'worthiness', commodification of vulnerable populations and capital accumulation strategies, and political and legal protections that pervade and are mobilized through these industries.

In a later section below I also reflect on the important though somewhat intractable methodological, ethnographical, and epistemological questions that my work raises. That is, how is it possible to assign or presume a subjective experience of violence and suffering and pain onto another being, whether human or nonhuman? To understand violence against human and nonhuman animals it is important to not only be able to epistemologically define what we mean by 'violence' and 'cruelty' in the first place, but also to be able to establish that suffering and pain are in fact experienced in carceral spaces such as the death house and slaughterhouse and not simply assume it to be so. And again, although the positionalities, subjectivities, and experiences of those exploitable and killable within these spaces are not the same, they do nonetheless share key aspects. Below I discuss various approaches to validating evidence, empirical and otherwise, of the emotional, psychological, and physiological trauma of inhabiting the body of the exploitable, disposable, and killable in carceral space.

Carceral logic and carceral space

Attempting to define 'the carceral' raises some basic questions about social and spatial relations. What makes a space a carceral space? What is it that makes us think of confinement of various kinds as being carceral or not? Is the carceral defined by the nature of the confinement – its intended purpose, or the inhabitants' experience? Does the intention behind confinement make a space carceral, or just the 'fact' of confinement? Where (if anywhere) does the carceral start and where does it stop? Is there an outside to the carceral and if so, where would it be (after Moran et al. 2017)? What are similarities and differences among various sorts of captivities, confinements, and incarcerations, particularly with respect to the physical spaces and practices that take place within them?

Moran (2015a) has defined carceral geography as a field of geographical research that focuses on practices of incarceration and zones of confinement, viewing 'carceral space' broadly as a type of institution with particular types of distributional geographies, and internal and external social and spatial relations. These include the architectures and geographies of carceral systems; the disciplinary regimes inherent in carceral settings; and the embodied experiences of imprisonment. Conventional wisdom suggests that the prison and jail are quintessential carceral spaces, with 'incarceration' conventionally understood as referring to the legal confinement of sentenced offenders under the jurisdiction of the state. What the carceral is then is related to what the prison is; it is anchored in the prison. So, as Bosworth (2010) has asked, "what is it about the prison that makes it a prison? What are its defining characteristics?" These are key questions; if the carceral is synonymous with the prison, it is a synonym that potentially surpasses the material and extends into the metaphorical (e.g. the 'prison of our minds'). Even within the U.S. criminal justice system, though, the carceral has also come to be understood as far exceeding imprisonment for criminal activity, embracing the myriad spaces and ways in which people are confined by the state, for example in the case of migrant detention (Moran et al. 2013; Mountz 2011; Loyd et al. 2012). Thus, the notion of spatial confinement and incapacitation is key to the carceral, but one that is expansive, that goes beyond the narrowly geographical to include a variety of practices, meanings, and social relations.

Carceral geography to date has concerned itself primarily with spaces of confinement broadly conceived through structural, political, and institutional contexts, operating at every scale from the personal to the global, and with a concern for everyday experiences and practices (Routley 2016; Moran et al. 2013; Loyd et al. 2012; Morin and Moran 2015). The carceral thus exceeds categories of criminality and penality, involving systems of confinement but which differ from those that a sociology of punishment or criminality would address. In this way, the carceral has come to encompass the spaces of detention of refugees, noncitizens, asylum seekers, the trafficked and the renditioned – as well as 'forms of confinement that burst internment structures and deliver carceral effects without physical immobilization', and embracing those trans-carceral spaces into which the more formally carceral constantly seeps (Moran et al. 2013: 240; Shabazz 2015b; Moran et al. 2017). Such spaces thus reflect the "carceral turn" and deployment of a new range of strategies of social control and coercion, with unprecedented fluidity between forms of confinement (Moran 2015a).

Davis (1990), for example, described Los Angeles as the quintessential "carceral city" comprised of heavily policed and privatized zones with accompanying segregated wastelands. Shabazz (2009: 285; 2015a; 2015b), Wacquant (2001; 2005; 2009), and others have proposed that the deployment of carceral techniques and mechanisms of prison punishment – surveillance, policing, containment, and restrictions on movement in homes, streets, housing projects, and neighborhoods, are the same as those used in prisons, thus effectively preparing especially young men for prison life. That many people inhabit trans-carceral spaces beyond the prison or jail is important to recognize, particularly with respect to the policing of Black neighborhoods and populations, who are then disproportionately represented in prisons and jails. Thus certain spaces and circumstances can *become* carceral through their very contact with the prison or other forms of actual custody (Orson 2012).

With respect to human populations in this study I focus (only) on the spaces, practices, technologies, and logics of the carceral as they relate to U.S. "correctional" institutions – prisons and the prison industry. Where my project extends beyond human corrections is with various nonhuman species and populations. In this book, I propose an engagement with interpretations of the carceral as including spaces and practices of nonhuman captivity, confinement, and enclosure, at various scales, be they state-sanctioned, quasi-legal, extra-legal, spatially fixed or mobile, and embodied (after Moran et al. 2017). Drawing these domains together into parallel conversation illuminates their shared carceral logics. Thus, in a broader sense my interests lie in defining carcerality in its structural/physical form; its functional mode of operation; its technologies and techniques; and in its experiential aspects. I study these carceral forms, modes, and logics within the Prison Industrial Complex (PIC), the Medical Industrial Complex (MIC), and the Agricultural Industrial Complex (AIC) (as well as other carceral sites such as the zoo). The carceral spaces and institutions within these 'industrial complexes' intersect quite fundamentally with respect to the violent experiential effects on the vulnerable populations within them. In drawing out a trans-species carcerality I argue that the carceral far surpasses spaces of human incapacitation to encompass the slaughterhouse, the research lab, the farm, and the zoo, and conceptually extends to the trans-carceral spaces and practices beyond these institutions proper (such as GPS tracking of wild animals). In that sense carcerality clearly exceeds categories of criminality, penality, punishment, and imprisonment.

Employing such an expansive definition of the carceral runs the risk of leading us to the unhelpful conclusion that just about any space could be considered carceral; i.e. that we live in a 'carceral age' surrounded by regimes of regulation and control that are aided by neoliberal reforms and the new security state apparatus, and these invade all aspects of life – they are simply 'everywhere' (Foucault 1977; Peck 2003). But if everything and everywhere is carceral, then the concept becomes evacuated of meaning (Bosworth and Kaufman 2011) and loses its potential for helping us to understand how a very specific carceral logic extends (only or mainly) to certain bodies and certain populations – and not to others – incapacitating and disposing of them in particular kinds ways and in particular kinds of spaces.

The view of a universal carceral owes to the foundational insights of Michel Foucault, particularly the last chapter of his *Discipline and Punish: The Birth of the Prison* entitled "The Carceral" (1977: 293–308). Here Foucault offered numerous terms and concepts related to the prison extended outward and that have been foundational to carceral studies generally, including: "carceral continuum," "carceral city," "carceral archipelago," "carceral network," "carceral methods," "carceral system," and "carceral texture of society." These are all a piece of and implicated in what Foucault described as "rippling carceral circles" emanating from the prison or prison-like spaces and reaching far beyond them, diffusing in carceral circles like ripples of water throughout society (1977: 298). Prison-like spaces such as almshouses and orphanages (what he called the "compact" carceral) extend in widening circles to charitable organizations and moral improvement societies (the "diffuse" carceral) that use carceral methods in ostensibly 'assisting' populations but also surveilling them. Beyond these lies the great carceral network of other domestic, urban, and embodied sites – the carceral archipelago – that transports the disciplinary techniques of the prison into the social body as a whole (1977: 298).

For Foucault, the prison was the centrifugal point from which the carceral circles radiated. With his model, the prison 'opens out to the social in its entirety'. On a number of levels Foucault's thesis of the universality of the carceral has been challenged, for example in its undeveloped spatiality (how far from the prison must the carceral circle extend before the influence of the prison is lost?) and inattention to the subjective experiences of carcerality.

Geographers such as Gill et al. (2016) address the spatiality question, outlining what they call carceral "circuitries" of people, objects, and practices that circulate in, through, between, and around carceral spaces in various ways. Rather than carving up the imagined geography of discrete carceral institutions they "emphasize their continuity ... the routes, courses, and pathways that constitute carceral space." I touch on such themes in subsequent chapters; for example, in outlining how the techniques and technologies of the slaughterhouse assembly line are notably similar to those of the prison execution chamber, owing to the circuitries of knowledge production and sharing across the U.S. and Europe in the 19th and 20th centuries (Chapter 2).

But where does Foucault leave us in terms of defining the carceral itself – with its subject matter so diffuse and universal making it almost impossible to define? Again, carceral geographers Moran et al. (2017) offer a concrete means of delimiting 'the carceral' to a set of three necessary conditions to be met in order for a space to be considered a carceral space. The first of these is what they call "detriment" – a space must inflict suffering, harm, or punishment, *experienced* as such and regardless of intent; the second condition is that of an agenic imposition of detriment via confinement (which would exclude, for example, voluntarily confining oneself to a gated community 'prison'); and the third, the material, virtual, or imagined space or spaces to which these relate. The carceral logic underlying this typology is useful for the examples I draw out in subsequent chapters. The logic of detriment and agenic confinement helps tie together the function, physicality, and practices of the spaces I discuss, even if 'carceral spatiality' might exist outside of the detrimental conditions just described. For example, as I discuss in the Methodology section below, violence takes many forms in the prison but nonetheless may not be experienced as such – and indeed may be experienced rather as somehow 'normal' and even subjectively beneficial and comforting. Thus, we must use caution when considering the nature of subjectivity itself for prisoners and animals in a diversity of carceral settings, recognizing that the embodied experience of the carceral is always a *relative* or *relational* one rather than an absolute one.

The above discussion begins to piece together the 'carceral logic' – the meanings, ideologies, and thinking – that underpin this project and that are contra to the rationalizations that give taken-for-granted credence to the same spaces in popular discourse and practice. Philosophers such as Gruen (2016a; forthcoming) focus on carceral logic as the logic of domination itself, shaping our social and political relations in order to naturalize domination and fix inclusions, exclusions, and disposabilities of certain bodies in the process. The carceral logic that allows us to call the zoo a prison offers a good example (Chapter 5). In such a carceral space, social ties of inside and outside are broken, and a type of disciplinary power spatially restricts bodies and manages them within highly orchestrated visible or 'invisible' structures, technologies, and practices. But in addition and quite foundationally, as already noted, within carceral spaces the bodies of the incarcerated are subjected as well to routine processes of 'animalization' while their status as property or 'person' is on-goingly negotiated. I continue this discussion of prisoner and animal bodies in the next section.

Incarcerated bodies: prisoners and animals

Geographers such as Buller (2014), Philo and Wilbert (2000), and Emel and Urbanik (2010) have brought into view human-nonhuman spaces and places that are mutually abjected and marginalized. Such work helps situate the historical and spatial contexts of lives and relationships that have been subjected to similar kinds of domination, 'othering', and violence. Philo (1998) has shown, for example, that humans working and living close to quarters of livestock slaughter are themselves seen as bestial and abject. A number of scholars have also examined shared animal and prisoner spaces at the prison (Haraway 2008: 63–65; Moran 2015b; Brower 2004). In Chapter 4 I touch on the co-labors of prisoners and animals on the 'inside' but this is not my focus. The sites I study are those wherein prisoners or animals are held captive and can become killable and put to death with impunity; can be exploited as medical and other testing material; can be subjugated for the entertainment of others; and can be commodified, treated as property and exploitable labor for the profit of others. Reading these case studies together enriches our understanding of the ways that humans and animals are living and dying in the service of capital, science, and putatively "safe streets" (Morin 2013).

I acknowledge that there are many connections that could be made between the spaces of confined or captive nonhuman animals and those of humans other than prisoners, produced through imperialism; gender, class, and racial difference; and economic exploitation (Nibert 2002; Patterson 2002; Cacho 2014; Kim 2015; 2016; Deckha 2010). Prisoners are just one among numerous abjected, marginalized, and animalized human groups – others including the enslaved, colonized, and displaced (Cacho 2014; Deckha 2010; Kim 2015). I also acknowledge that using the term 'nonhuman animal' masks the vast differences among nonhuman species; reproduces the arbitrary and socially constructed human–animal divide, since humans themselves are animals; and also casts *non*human individuals as existing in a state of lacking something ostensibly superior. Nevertheless, in this work I focus on certain types of sentient land animals (identified by species) who, evidence shows, are among those able to suffer, to feel anxiety, fear, and pain.

Obviously, though, there are difficulties in attempting to determine some sort of 'dividing line' or boundary among living creatures to which my argument could be conceptually extended – those who have the ability to think and reason versus those who do not; those who have the ability to move and respond autonomously versus those who do not; and so on (e.g. Hudson 2011: 1661). Are honeybees and other insects, animals? Should animals who are not transformed into meat and other commodities for human use also be taken into consideration? What is animal or nonhuman clearly varies over time and place, and associated ontological assumptions have enormous ethical and other implications. For purposes of my analysis, however, it is not necessary to establish such a dividing line in order to call attention to some of the most egregious examples of infliction of pain and suffering on human and nonhuman animals through carcerality and captivity.

I call attention to some examples acknowledging that there are many others, which a number of geographical studies have addressed (e.g. Srinivasan 2013; Gillespie 2014; Gillespie and Collard 2015; Braverman 2011; Nast 2015). Criminologists have begun calling attention to these links as well. Taylor (2011: 252–253), for example, explains that animals never enter criminology studies as agenic subjects, but only as objects – such as in studies of individual humans accused of cruelty to animals. Criminology is just beginning to consider taken-for-granted, large-scale normalized animal abuses as 'criminal' – even if they are not considered as such in the dominant culture because they are not recognized as 'criminal acts as legally defined' and because animals do not (yet) have legal standing in the courts (but see Deckha 2013a; 2013b and discussion in Chapter 4).

While we may observe equal vulnerability and systems of abuse within human and nonhuman spaces of captivity, "susceptibility to incarceration" (Acampora, 2006: 108) itself might remain invisible to our conscience, particularly with respect to the structural racism that is a basic constituent part of the U.S. Prison Industrial Complex. I discuss this at some length in Chapter 3. As noted above, in this study the carceral spaces that humans inhabit are exclusively those within the U.S. correctional system. Many aspects of prison life and the prison setting inform my study, including those that Moran (2015a) has identified as subjects of carceral geography, including that of occupying and moving through prison buildings and the emotional and embodied geographies of prison life and experience. Within these carceral settings I have selected several representative sites for study, placing them within their respective historical and geographical contexts: prisoners' 'killability' within death row, and their exploitability and disposability within factories and other sites of work programs, medical labs on the inside, and solitary confinement cells.

Captives' first-hand accounts of their experiences provide a necessary dimension to this book – from the accused's description of the execution chamber and events on the last day to the bleating of animals prodded and stunned to unknown futures – all provide important evidence not only of the emotional, psychological, and physiological trauma and violence of inhabiting the body of the exploitable, the disposable, and the killable, but also provide important insights into the physical spaces and technological and other control features surrounding the spaces of torture and death. I would underline again, though, that I do not assert that the human and nonhuman embodied experiences within sites of slaughter, experimentation, entertainment, and labor are the same, nor that the positionalities of the exploited subjects are the same; they are not, although they share key aspects.

The day-to-day embodied experience of captivity and incarceration, of being identified with a number, a tattoo, a brand, and other forms of bodily modification; the emotional and psychological strain of knowing the approach of death or of the testing apparatus or whip, all are interwoven into the day-to-day carceral spaces of the prison, the lab, the farm. Gruen (2014b) argues that it is the loss of dignity and respect that is the most insidious outcome of carceral captivity and thus should be the main point of a carceral comparison, as human and nonhuman bodies are subjected to both physical controls and constant surveillance. These institutions of enforced occupancy, violence, and control "ensure the production of docile bodies (or dead ones)" (Acampora 2006: 108–110). Moreover, if ever released, such humans and nonhumans would be subjected to their respective carceral milieu: electronic tracking devices used to track animals are not unlike the parole boards (or degrading tracking devices) of released prisoners.

As noted above, among its most detrimental effects, 'the human' and 'the animal' themselves are made and unmade through encounters with carceral spaces. Said another way, carceral sites and institutions reinforce understandings and perceptions of the ontological status and identity of the animality of beings, and in turn the process of animalization itself subjugates both certain human and certain nonhumans into hierarchies of worthiness and value. As Kim (2016: 38) provocatively shows via the example of the shooting of Harambe the gorilla at the Cincinnati Zoo after a child had fallen into his enclosure, zookeepers are above all the "keepers of humanity" (Chapter 5). It is this carceral logic of domination, in concert with that of animalization, racialization, and criminalization of human and nonhuman bodies that is contrived, reinforced, and becomes fixed in carceral spaces.

I recognize though that the politics and ethics of making comparisons between racialized and classed human lives and that of nonhuman animals in respective carceral spaces can be problematic and fraught. It is challenging for humans who are embedded in violent, racialized, and criminalized human histories and spaces to not be offended by posthumanist comparisons to animal suffering (Kim 2010; 2011; Deckha 2013b; Jackson 2015; Ko 2016). As noted above, the category of 'human' is contested in any case, and it is important to not move too quickly 'beyond the human' without acknowledging the continued exclusion of many human lives from full incorporation within it. And yet thinking particularly about race and animals together is important, precisely because of the way that racialized people have been and continue to be animalized in carceral spaces (Chapter 3). Moreover, the carceral logics of domination are intertwined across human and nonhuman groups. To take one more example, as Deckha (2013b) has shown, animal anti-cruelty legislation has the double effect of selecting certain animals for protection while targeting the behaviors of certain minoritized populations of people as deviant and transgressive (also see Elder et al. 1998; Kim 2015). Meanwhile, industrial practices involving the dominant culture – as well as the abuse and killing of most animals – remain immune from critique.

When considering the structures and industrial regimes that allow for incarcerated bodies to be exploited for food, experimentation, entertainment, or labor, it is useful also to point out the different kinds of 'deaths' imposed upon them in the process. Prisoners and animals confined to carceral spaces live lives that are threatened with both biological death and with 'social death'. In the latter case, they become part of what we might call the 'living dead'; an individual can be biologically alive but through captivity, be completely unmoored from social life and conditions that make for a live-able life (Butler 2009: 21–23; Agamben 1998; Guenther 2013; Mbembe 2003; Patterson 1982). As Butler has argued (2009: 1–4), if certain lives do not qualify as lives or are, from the start, 'not conceivable as lives within certain epistemological frames', then these lives are never considered to have lived nor be lost in the full sense – they are not "grievable." These are lives in a constant state of precarity (Butler 2004: 5). Specific mechanisms of power generate specific ontologies of subjects that regulate this association, and as I argue throughout, the processes of animalization, racialization, and criminalization are several of these foundational mechanisms operating in carceral spaces.

Animal 'criminality'

The nonhuman animals in the spaces I discuss are of course not abjected as 'criminals' to be incarcerated and punished; yet, as property of humans and through various other relationships with them, they become functionally a part of the carceral. As Taylor (2011: 252–253) explains, animals have traditionally entered the remit of criminological studies only as objects, and within what is legally defined as "criminal" (above). Historically, however, it is worth noting that certain animals have been treated as criminals, demonstrating agency, put on trial, prosecuted, and executed for their 'crimes' (Evans 1906; Beirne 1994; 2009; 2011; Girgen 2003; Dinzelbacher 2002). When considering the criminalization of nonhumans, it is important to reflect on when, where, and why such practices occurred and then subsequently ceased (if indeed they have). This seems especially worthwhile with respect to assigning moral and judicial guilt to animals, as well as to assumptions made about their understanding of the sentences imposed. In addition, we might ponder what this process of criminalization (and execution) of animals says about humanness itself – particularly the humanness of nonhuman animals – and thus inter-species boundary construction.

Within the European context, Evans (1906) examined a wide range of medieval animal crimes and trials, distinguishing them by those prosecuted via ecclesiastical courts versus those by secular courts (the former responsible for prosecution of offending animals who had been domesticated by humans, while the latter issued against wild animals, plagues, vermin, and pests). Animals put on trial and prosecuted ranged from insects and worms to mice, donkeys, horses, bulls, pigs, dogs, wolves, primates, and dolphins, "their crimes rang[ing] from impersonating a cleric to murder and to causing droughts and famines" (Evans 1906; Beirne 2011: 361; Girgen 2003; Siebert 2014). Among many historical examples, in 1275 a monkey was found guilty of espionage in the English Channel. As Dinzelbacher (2002: 420–421) maintains, animals could be and were considered human-like, with the 'gap' between them beginning to narrow in the 12th century: "[alongside] the numerous animal epics that presented the animal kingdom as a mirror of human virtues or vices … the animal trials undoubtedly betray the same tendency to reduce the ontological distance between man and beast." Girgen (2003: 115–120) argues that it would have been much easier to simply kill the animals during these periods rather than put them on trial, but biblical notions that animals should be held accountable for their actions dominated. Such notions were founded on both a 'search for lawful order' in the universe as well as simple retribution.

Perhaps unbeknownst to most, such practices continued well into the 20th-century United States. In the first decades of the 20th century animals were tried and punished for crimes in Kentucky and Tennessee, and aside from various 'folklore' tales (such as that of a dog sentenced to life in the Philadelphia State Penitentiary for killing the governor's cat), serious punishments were enacted on animals for crimes committed. Perhaps the most poignant examples are those of elephants destroyed within the carceral spaces of the circus industry: 'Murderous Mary' was hanged for killing her trainer in Tennessee (Burton 1971), and Topsy the elephant was electrocuted at New York's Coney Island circus in 1903 for a similar crime. After an unsuccessful attempt at hanging Topsy, she was given cyanide and then Thomas Edison, in competition for creating the strongest electrical current, oversaw her execution on an electrified platform.

While these examples may seem of another age and sensibility, Girgen (2003: 122–128) argues that holding animals responsible for criminal action continues today, for example in punishing and killing 'vicious or dangerous dogs' (also see Nast 2015; Glick 2013). The difference today, he astutely notes, is in the enactment of 'summary justice' against the animals: unlike historically, no formal charges are levied, no legal counsel is provided, no public hearings are held, and no due process is awarded the accused; and moreover, executions now typically take place in the private space of a euthanizing veterinary office or animal shelter rather than in the public square. Girgen (2003: 129–130) moreover asserts that the reasons for executing animals for crimes today seem to be revenge, or for 'restoring the hierarchical order' that the animal upset. Consider the summary killing of the gorilla Harambe at the Cincinnati Zoo in 2016 (above) – not for something he did but rather for something he might do (i.e. harm the child who had fallen into his enclosure; Gruen 2016; Kim 2016). I find such examples helpful in discerning just how fluid meanings of 'the criminal' and 'the carceral' can be across time and place. They offer a useful segue to an overview of those industrial-scale institutions that foundationally comprise the U.S. carceral landscape of today.

Carceral entanglements: The Prison-, Agricultural-, and Medical-Industrial Complexes

As Best et al. (2011) and others have argued, the late 20th–early 21st-century historical epoch can be best characterized as one large pervasive, expansive, and insidious 'global industrial complex' that combines the logics of capitalist exploitation and profit with industrialist norms of efficiency, control, and mass production. This naming was first applied to the Military Industrial Complex, depicting in the mid-20th century the ever-closer relationship between the U.S. military and defense companies, contractors, and manufacturers who were set to make millions (and indeed billions) of dollars developing political interests abroad along with the tools necessary for waging war – "war for profit" (MIC 2017). Similar forms and structures, modes of operation, and violent effects characterize the trans-carceral milieu that extends through the Agricultural-, Medical-, and Prison-Industrial complexes. In this project I tease out such relationships in these three industrial complexes for the purpose of exposing the ways in which they exploit and enact violence on prisoners and animals. Their sites are collectively the locations where vulnerable human and nonhuman bodies suffer and become, as Collard and Gillespie (2015: 4) express it, "sellable, buyable, breedable, displayable, and eventually, killable."

The developmental similarities across the Agricultural Industrial Complex (AIC) and the Prison Industrial Complex (PIC) are particularly noteworthy. Both rapidly expanded during the 1970s and 1980s era of deregulated 'big agriculture' on the one side, and changes in drug and sentencing laws that led to new thresholds in mass incarceration on the other. During the Reagan era of the 1980s, new forms of large-scale, mechanized agriculture emerged alongside deregulation of the agricultural sector and new subsidies for farmers, all of which gave rise to unprecedented expansion of certain forms of mass-produced food commodities. Meanwhile federal and state legislation associated with the War on Drugs (particularly focused on crack cocaine; Alexander 2012), alongside other 'tough on crime' and 'law and order' campaigns such as the federal government's 1984 Sentencing Reform Act and California's 'three strikes' rule, resulted in unprecedented numbers of people behind bars. In 2014 that number peaked at 2.4 million, although it has decreased slightly since, primarily with the release of non-violent drug offenders incarcerated in federal prisons. Yet if one were to count all U.S. adults trapped in the PIC – those imprisoned as well as those on probation or parole – that number skyrockets to 6.7 million (Williams 2016).

The Prison Industrial Complex (PIC) itself is shorthand for the rapid expansion of the U.S. prisoner population and the political influence of private prison companies such as Corrections Corporation of American (CCA) that supply goods and services to what are, after all, government agencies that are supported by local, state, and federal taxpayers. Davis (1995) coined the term *Prison Industrial Complex* to signal commodification of 'prisoners for profit'. As Peck succinctly observes, "this is not less government, but different government. This is a more punitive approach to social marginality," an "uneasy marriage between economic liberalization and authoritarian governance" (2003: 225). The reduction in the welfare state is used to legitimate and harden new regulatory regimes and new forms of governmental rationality, suggesting that "the prison system can be understood as one of the epicentral institutions of these neoliberal times" (Peck 2003: 226; also see Pratt et al. 2005; Harcourt 2010).

Thus, the PIC most fundamentally describes the overlapping and intertwined interests of government and industry – which use surveillance, policing, and imprisonment as 'solutions' to economic, social, and political problems – keeping as many people locked up as possible. The private prison and its ancillary private companies such as CCA today earn unprecedented profits on the bodies of the incarcerated through control of individual sectors of the PIC such as in health care and food service delivery, and most recently, in handling the 'business' side of probation and parole. Such companies include everything from corporations that contract prisoner labor, construction companies, surveillance technology vendors, companies that operate food services and medical facilities, and private probation companies, lawyers, and lobby groups (McCormack 2012; Palaez 2014; Critical Resistance 2017). In short, the PIC includes a vast array and integration of jobs, private contracting of all sorts of activities and supports, health care, food, all which expand beyond the prison itself into such arenas as the policing of neighborhoods.

The U.S. Agricultural Industrial Complex (AIC) represents a shift from traditional (small farm) modes of agriculture to industrial-style production that is now global in reach. Industrial farms are very large, highly specialized, and run like factories with heavy inputs from machinery, fossil fuels, water, pesticides and other chemicals, and synthetic fertilizers derived from oil (Union of Concerned Scientists 2016). The AIC is also characterized by monoculture cultivation of a single crop for food, feed, fiber, or fuel purposes. The same type of relationship exists with the PIC in respect to relations between government, business, and the agriculture industry. Government support of industrial agriculture – particularly with respect to meat and dairy production and concentrated feeding operations (CAFOs), has been supported by 30+ years of U.S. farm bills that have provided subsidies to operations that provide food at a low(er) cost to consumers but which also damage the environment, use excessive energies, and which of course rely on purpose breeding of chickens, cattle, pigs, and other animals to unprecedented degrees (see Chapter 2).

As Thomas and Shields observe (2012: 4), the nexus of mass incarceration and mass exploitation of nonhuman animals is unprecedented in history, each developed along similar lines of carceral logic:

> The connection between animal studies and incarceration discourses has never been more intimately associated … [we note] the eerily similar trajectories of the prison industrial complex and factory farms. Both institutions developed rapidly, sprouting up in rural areas and proponents for both heralded them as job-providers for impoverished communities. Both institutions serve as transformative spaces that encourage physical displacement, limit mobility and create exiled individuals. Both institutions forge identities, shape relationships and take lives. State-sanctioned killings are capital punishment in one arena and "processing" in another.

Merritt and Hurley (2014) were among the first scholars of whom I was aware to compare the development of the AIC and the PIC, pointing out in a paper presented at an Association of American Geographers meeting the parallel rise of mass incarceration and CAFOs in the U.S. within the last four or five decades. They outlined some ways in which these industries share spatial design, disciplinary technologies and practices, and objectified and terrorized bodies as their 'products'. Merritt and Hurley also argued that the systematic and sustained violence of the PIC and CAFO commodify animal and human bodies in similar ways through confinement, restriction, and pain.

Following on their work and fundamental to my analysis is the fact that certain geographies of the PIC and AIC (as well as the Medical Industrial Complex, below) are similar – in their location, spatial layout and design, and in the mechanisms and technologies that regulate movements within them. In the space of the auction block, the slaughterhouse, the sterile laboratory, and the execution chamber, all movement is carefully choreographed, routinized, surveilled, and controlled. Among other things these methods speak to the perverse underlying notion of the 'precision of the correct death', however much the putatively 'advanced' science or technology intrinsic to them is but a façade (e.g. Basu 2015). Such operations even look similar in their rectangular building designs, muted color schemes, surveillance mechanisms, and locational 'advantages' in remote and rural areas. The distribution of carceral sites in rural, remote, or secreted locations connotes spaces 'hidden in plain view' – so innocuous and ordinary are their color and architectures that they do not attract attention (Merritt and Hurley 2014; Pachirat 2011).

As Emel and Urbanik (2010: 208) argue with respect to the factory farm (but which could also be said of the execution chamber and pharmaceutical testing lab), geographers have made important contributions to the study of the ethics of such spaces precisely by paying attention to the invisibility of their locations (e.g. Elder et al. 1998). Pachirat (2011), too, convincingly argues that society's most violent processes require them to be hidden from view, even from those who are actively engaged in the violence (the slaughterhouse workers, lab researchers, the executioners on death row); their participation requires a deep compartmentalization and desensitization, as well as a spatial organization of the site such that the violence is so thoroughly routinized as to be invisible even to them. I engage such arguments in Chapter 2, considering further relationships between sight and (dis)avowal of such practices.

Turning to the Medical Industrial Complex (MIC), it too shares many spatial, structural, operational, and effective outcomes of the PIC and AIC, as well as sharing many of the same types of relationships across government and industry. Testing of drugs and other products is one among a host of activities, practices, and professions that are part of the much larger 'health industry' and apparatus of the MIC – including but not limited to doctors and medical schools, hospitals, nursing homes, insurance companies, drug manufacturers, hospital supply and equipment companies, real estate and construction businesses, health systems consulting and accounting firms, attorneys who file medical malpractice suits, manufacturers, retail pharmacies, and banks. As the Ehrenreichs (1971) stated it, the concept of the Medical Industrial Complex (or "American Health Empire") conveys the idea that an important, if not the primary, function of the health care system in the United States is business – that is, to make profits (Dober 2008; MIC 2017).

The MIC includes a panoply of institutions and actors that are not part of my analysis; I focus only on "Big Pharma," the pharmaceutical industry surrounding, supporting, and benefiting from the development of drugs and other products, that share a vast network of scientists and laboratories located at research universities, federal government agencies such as the National Institute of Health, and other public and private research companies and facilities. I discuss the phenomenal scale of the Big Pharma industry in Chapter 3, its profits and ostensible benefits to human society, alongside the phenomenal cost to 'animalized' lives, human and nonhuman, made exploitable and disposable in routine laboratory research and experimentation in the U.S. each year (Hornblum 1998; Washington 2006; DeMello 2014; Urbanik 2012; Ross 2014).

Ultimately I study the Prison-, Agricultural-, and Medical-Industrial Complexes (along with a key site of the Entertainment Industrial Complex, the zoo; Chapter 5) together because the similarities across them are so pronounced as to warrant a productive carceral comparison. Their common denominators allow us to link their processes and profits, and in so doing, ultimately challenge the social and spatial norms that produce and shape the trans-species 'carceral continuum' (after Foucault 1977: 297). Before concluding with a chapter-by-chapter outline of how I engage these 'industrial complexes' at specific carceral sites and institutions, I pause next to consider some of the crucial methodological issues and challenges that this work presents.

Methodological considerations

One of the assertions of this book is that carceral geography and critical animal studies intersect at the nexus of humans' and animals' lived experiences of suffering. But taking a step back, if we are to construct a meaningful and 'reliable' methodology to study and apprehend prisoner and animal experiences, what would it look like? The methodological challenges Buller (2015) and Carter and Charles (2011: 9–27) identify relate especially to ethnographical study of animals – how do we know animals suffer, feel pain, grieve, and so on, as well as how do we understand their resistance (response) and/or agency with respect to their conditions of confinement and captivity (Gillespie 2012; Wadiwel 2015; Kim 2016; Haraway 2008; Derrida 2002)?

Can we ever know what animals think and feel? How do we not 'silence' the non-linguistic being? Bentham's oft-quoted philosophical proposition about animals (1781: 4), "The question is not, can they *reason*? Nor, can they *talk*? But, can they *suffer*?" is as relevant today as ever. And yet it is also the case that forming an ethnographical and epistemological understanding of human prisoner suffering and pain – or indeed, defining what might be considered cruelty and violence to prisoners – can be just as fraught and methodologically challenging. How can we ever 'know' the suffering and pain of another being? Are such experiences (only) culturally or individually specific, or can we establish some objectively harmful 'detrimental' parameters produced in carceral settings, for both prisoners and animals (cf. Moran et al. 2017; Crewe 2017)? Among other key questions is how to address the comparability issue, the comparability of suffering of very different subjectivities of oppressed beings in their respective carceral spaces.

Many presumptions of suffering and pain underlie my discussion about human and nonhuman experiences such as in the prison execution chamber and slaughterhouse (Chapter 2). Yet in order to understand violence against humans and animals, it is important to be able to establish, from the point of view of the subjects, that they are indeed suffering and experiencing violence and not just assume it to be so. Such becomes a poignant question whenever we attempt to understand another human when power relations pervade the relationship unevenly, but this is particularly tricky when considering the power relations that pervade the human–animal relation when we cannot avoid intellectualizing, mediating, interpreting, and filtering what it is we assume to know (Buller 2015). Thus, the methodology questions are important to address: how do we understand animal experiences and 'speak for' them in our work, or any other 'other' whose interpretation of various acts and experiences may differ from our own in ways too innumerable to mention (e.g. Mowe 2016)? As Urbanik (2012: 183) adds, "what methodologically does it take to study the subjectivities of species as diverse [as] a lemur, a hammerhead shark, a hummingbird, and a tiger?" Various ethnographic methods have been tried, ranging from multi-species ethnographies to single species ethnographies, ethological and behavioral observation, participant observation, interviews, archival methods, quantitative and statistical methods, tracking procedures, mapping, and visual and visceral methods (after Colombino and Steinkrueger 2015). In the remainder of this section I examine the potentiality of some of these methods, as well as consider how we might think about what constitutes 'suffering' or 'violence' in the first instance for both human and nonhumans within carceral settings.

Suffering and the animal other

> The key methodological endeavor of human–animal relational studies has been to come to some emergent knowing of non-humans: their meaning (both materially and semiotically); their 'impact' on, or even co-production, of our own practices and spaces; and our practical and ethical interaction with and/or relationship to them – or at least to find creative ways to work around unknowing.
>
> Henry Buller (2015: 379)

Buller (2015: 374–375) addresses methodological issues within animal studies by asking both what do we know of animals, and what might we do with that knowing? Obviously animals express themselves. He observes that because nonhuman animals do not 'speak' or 'talk' they have long remained, fundamentally, nature's silent objects to human subjects (although as Smith [2002: 54–55] adds, the question of how individuals speak is clearly a social rather than an ontological matter; consider the vast differences in forms of communication, across species but also within them, including within human groups). Buller describes the historical observational, 'mechanistic' methods of animal study within the natural and behavioral sciences that rely on human representations of animal experience, arguing that we must find alternative ways to allow animals to speak and represent themselves. Hodgetts and Lorimer (2015) and Kirksey and Helmreich (2010), for example, discuss developing tools and methods that (at least arguably) "leave out [interpretations of] the gatekeeper," such as monitoring and tracking devices. Griffin (1992), in discussing his cognitive ethological (vs. behavioral ethological) method, asserts that we must start with the proposition that animals are sentient and conscious, albeit with fundamentally different types of experiences and consciousness than humans. Such non-anthropomorphic, multi-species ethnographic methods accept the conditions of 'mindful animal agency in human–animal relations', in which the exchanges between humans and nonhumans are practiced and performed between 'subjects-in-interaction' who have 'shared biographies' (Buller 2015: 377; Davies 2013; Emel et al. 2002; Law 2012).

Haraway (2008: 205–231) has argued that there are "flourishing, generous, worldly, and revealing" shared languages in moments of "fleshy interspecies interaction beyond calculation than the chasm of linguistic difference" would otherwise dictate. Livestock in particular, Buller observes, might offer distinctively visceral, performative, and affective opportunities for exploring co-presence and mutual becoming within the context of animal welfare (2015: 379–380), particularly in ethological studies of animals' perceptions within spaces of confinement, from shelters and zoos to farms and laboratories. Intra- and inter-species communication has been a focus of such ethological research – and importantly, as an alternative to the visual bias of most observational methods (Hodgetts and Lorimer 2015: 288–289; also see Hemsworth 2015):

> This communication tends to be aural, consisting of calls, barks, whoops, growls and the like; and researchers are able to develop a degree of understanding through investigating the consistency of aural communications between individuals of an animal species in relation to external events.

Many scholars argue for more such visceral and affective approaches to the study of animal experiences. Gruen (2011: 114–115) considers nonhuman animal pain as being both a sensory and affective/emotional experience. She argues that there are two objective, or at least observable, ways to identify pain: one is physiological and the other behavioral. While physiological studies track brain stimulus and response, behavioral evidence of animal pain is strong:

animals guarding a particular area of the body, neglecting grooming, altering facial expressions, changing sleep patterns and social interactions, vocalizing in unusual ways, licking, biting, scratching or rubbing areas of the body, panting, sweating, and refusing food and water all constitute such evidence. She asserts that most animals experience conscious pain, despite differences in pain thresholds (in humans as well as animals), and the fact that animals can perhaps consciously hide pain in order not to feel vulnerable. Paying attention to these behaviors is to Gruen (2015a) "caring perception," a blend of emotion and cognition that is based on knowledge and recognition that we are in deep relationships with other beings.

Thus, we might consider as highly compelling the more visceral, non-representational, emotional, affective approaches to the study of animal suffering; 'visceral knowing'. Buller (2015: 378) asserts that humans "may not share language with non-humans but we do share embodied life and movement and, in doing so, different – yet both biologically and socially related – ways of inhabiting the world." As Barad (2003) asserts, 'we know' because 'we' are of the world too. We learn 'by witnessing'; if the phenomenological experiment of encounter is pushed far enough, a portrait of shared existence emerges (Lorimer 2010: 378–379). Dayan (2016) and King (2013) similarly argue for such an empirical approach; that because we come to see and know through our senses, we should privilege the 'primitive', the pre-linguistic as the place of 'mutual habitation' between humans and animals.

In writing about animal grief, King (2013: 6–8) adds, though, that while animals feel sorrow "goat grief ... is not chicken grief. And chicken grief is not chimpanzee grief or elephant grief or human grief. The differences matter." King's is an important insight, not because any one type of grief is more consequential or meaningful, but because the problematical human–animal divide can be reinforced when we say that forms of pain and suffering are the same or similar across species – with those who are perceived as more 'human-like' taken more seriously. Meanwhile, those nonhumans whose emotional capacities or cognitive abilities are not 'like us', can become even more vulnerable to violence, exploitation, and killability.

In attempting to interpret the subjective experiences of the animal 'other', I also find applicable the more familiar (to me) terrain of colonial and post-colonial theories and methods. Haraway (1989: 10–11) provocatively made the allusion to Edward Said's *Orientalism* (1978) with her "Simian Orientalism," suggesting that the field of primatology constructed the metropolitan (cultural, human) self, if you will, inexorably from a relation with the animal (natural, peripheral) 'other'. Questions of representation and resistance appear foundationally in colonial and postcolonial studies. And although animals do not appear as 'colonized subjects' of humans within this work (beyond their domestication), they also are not 'passive others' under the human gaze despite relationships that are replete with inequality, marginalization, essentialism, oppression, and resistance. As Kirksey and Helmreich (2010: 554) observe, after Appadurai (1988: 17), this is the canonical problem faced by anthropologists, "the problem of voice ('speaking for' and 'speaking to')." Deckha (2010: 45) usefully adds that this thinking is pervasive throughout Western imperialism wherein 'the other' that the 'autonomous liberal actor' differentiates himself against is always heavily reliant on the category of the lesser subhuman or nonhuman.

Reviving Pratt's (1992) theory of "contact zones" offers a way out of this conundrum. She argues for a 'transculturation' process, one in which encounters among speakers of different languages seek to define the 'contested' zone in their own way (also see Haraway 2008: 205–231). As Pratt (1992: 6–7) articulates, a "contact perspective emphasizes how subjects are constituted in and by their relations to each other …. It treats the relations in terms of co-presence, interaction, interlocking understandings and practices, often within radically asymmetrical relations of power." Examining the material practices that mediate interactions in any given contact zone offers clues about their meaning to the respective participants (a method I have applied in various previous works, e.g. Morin 2008). Certainly within the contact zones of animal slaughter, for example, paying attention to the material practice of forcing livestock through trucks, cages, pens, and chutes offers innumerable clues about the meaning of the encounter to the animals: they categorically attempt to escape from – resist – these spaces at every opportunity if not for the electric prods and other pain-inducing equipment that prevents them from doing so (Smith 2002; Higgin et al. 2011; Gillespie and Collard 2015; Gillespie 2014; Pachirat 2011; Gruen 2014; Rasmussen 2015). This, despite what Smith (2002: 50) describes as the physical (as well as linguistic, emotional, and ethical) distancing practices that humans employ to limit livestock's scope for self-expression.

Grandin (2012, above), a spokesperson popular with the global meat industry, employs a 'somatic sensibility' in designing what she considers humane slaughterhouses that reduce the stress, anxiety, and suffering of the animals passing through them. Animals' voices can be stilled by mechanisms such as use of indirect lighting, non-slip floors, curved corrals that visually block what is coming next on the assembly-line, and hold-down restraining covers on conveyors. By becoming sensitive to animal affectivities and energies, such innovations promote the notion that 'humane killing' is possible since animals appear calm, quiet, and nonresistant during processing (Greenhough and Roe 2010: 44; Smith 2002: 56). But stifling animal self-expression in these ways also suggests simply more calculated means of making the killing easier on the killer, bypassing ethical questions altogether (cf. Patterson 2002). Womack (2013: 12–13) provocatively argues that (even) within Native American beliefs and practices, "there is no respectful way to kill an animal":

> If somebody shoots me with a high-powered rifle, I'm not going to like it no matter how many prayers and ceremonies the guy does before he pulls the trigger …. The prayers and ceremonies do something for us, not the deer, at the very least not the same thing for the deer, and there is no way to escape the fundamental inequity of the relationship. I would go as far as to say the lack of relationship: she's dead, we're not.

Epistemologies of prisoner and animal violence

The above discussion prompts us to question whether there exists any absolute, normative conditions and practices that constitute cruelty and violence to humans or nonhumans. With respect to the embodied experience of being subjected to surveillance, regulation, torture, and death, how can such practices be classified as violent and cruel to 'others' who may speak a different language or may perhaps experience acts and activities in different ways? Moran et al. (2017, above) suggest that in order for a space to be considered carceral it must inflict 'detriment' (suffering, harm, or punishment), *and* be experienced as such by the subject, regardless of the specific intent. But is it sufficient to assert that "the carceral is [only] in the eye of the beholder?" As Moran et al. assert, "what is felt acutely as suffering by one individual may not perturb another … what is not intended to punish, on the other hand, may deliver significant harm." Such alerts us to the fact that there are obvious risks involved in both avoiding or asserting normative definitions of cruelty and violence outside of those experiencing them. But how do we proceed then, respectfully recognizing that the nature of subjectivity itself varies tremendously within various carceral settings? One plausible way to detect or define cruelty and violence is to take seriously the affective resistance to the conditions of confinement on the part of the incarcerated subject (e.g. Gruen 2011). But what of those subjects who have come to consider their conditions 'normal' and do not resist?

What determines 'animal cruelty' has of course been impacted historically by a wide variety of human interests that are contextually fluid, with much of it, of course, legally sanctioned. Understanding what comes to be defined as animal cruelty necessitates a broad and deep analysis of the sociological, economic, and legal forces in play in a given society at a given time.

The attribution of savagery itself is place- and culture-specific: the same act or practice – and those engaging in it – could be considered savage and cruel in one context while perfectly acceptable in another. Kim (2010: 61–62) asserts that defining cruelty is a cultural matter and that many 'racist double standards' are often in play when attempting to determine universalist, normative claims about it (also see Nast 2015; Deckha 2013b, above). How then might one respect or accept practices of others that might seem abhorrent, yet retain an ability to name and disavow any particular practice as cruel? Casal (2003) suggests that cruel practices should not be favored with legal protections just because they are part of religious or cultural traditions. Kim (2010: 64–70) proposes that when considering 'relative' versus 'universal' cruelties, it is hierarchical and dualistic thinking that gets in the way. To her, we need to find less hierarchical ways of relating to other beings, without assuming that any single form of domination is more foundational than any other. Kim (2015: 16–18) suggests thinking in terms of taxonomies rather than dualisms and hierarchies to ask questions about "relationality, positionality, and multidimensionality" that are obscured by both the human–animal dyad as well as hierarchies that humans ascribe to various animal species that determine their relative worthiness – e.g. which are 'good', to be protected and admired, versus those that are threatening or worthless. Once we develop a better understanding of how animal lives come to 'matter', the nature of cruel, violent, or abhorrent acts done to them becomes clearer.

Measuring cruelty and violence within the prison presents its own significant challenges of interpretation. Most obviously, official methodologies for reporting 'violence' – most typically assault rates, either prisoner–prisoner or prisoner–staff – are suspect; accurate data reporting varies from state to state and among different types of institutional cultures. Most institutions record only those incidents that reach a particular threshold of seriousness, such as those that require medical care or hospitalization. Even when attempting to limit analysis to physical assaults only (Bales and Miller 2012), accurate records on these can be difficult to acquire since official estimates grossly underrepresent physical violence (for obvious reasons), and there is strong disincentive for prisoners themselves to report violence due to fear of reprisal or retaliation. On top of this, not all incidents of violence are perceived as such by prisoners, and some prisoners are clearly more victimized than others (Wolff et al. 2007: 596; Crewe et al. 2014: 57).

It is important to keep in mind that prison cruelty and violence can take many forms – from mental abuse and racial slurs to sexual predation and rape, theft and property damage, ubiquitous systems of coercion, constant surveillance, stimuli deprivation, lack of privacy, crowding, excessive noise, the 'violence' of resignation and despair, perceived threats to bodily harm – to actual beatings, stabbings, and other types of physical assault. Prisoners on death row (Chapter 2) spend 23 or 24 hours a day in their cells; they eat, sleep, live, and exercise alone in their cells, a practice that is designed as pure torture and punishment even without the added stress of anticipating execution (Mulvey-Roberts 2007; Guenther 2013; Rhodes 2009; Morin 2013).

The oppressive day-to-day norm is that of sensory deprivation, isolation and loneliness, enforced idleness and inactivity, oppressive security and surveillance procedures, and despair. "Prisoners subjected to prolonged isolation may experience depression, despair, anxiety, rage, claustrophobia, hallucinations, problems with impulse control, and an impaired ability to think, concentrate, or remember" (Magnani and Wray 2006: 100). Some prisoners have become so disoriented and out of touch that they "act out to prove they still exist" (King et al. 2008: 161; Guenther 2013). The mental illnesses associated with and created by isolation and idleness, and the corresponding lack of psychiatric care available to prisoners, lead many psychologists and criminal justice scholars to declare such practices and sites without question, "cruel and unusual" punishment (Dayan 2007; Arrigo and Bullock 2008; Cloyes et al. 2006; Martel 2006; King et al. 2008). These offer important evidence and methods for conceptualizing cruelty, torture, suffering, and pain to death row prisoners, and yet in the end we need little more evidence than that provided by the autobiographical writings of prisoners themselves (Vickers 2014; *Witness to an Execution* 2000; James 2005; Mulvey-Roberts 2007).

I take up these issues in subsequent chapters with respect to prisoners on death row (Chapter 2) and those restricted to solitary confinement (Chapter 5); as well as with respect to the ambiguous nature of prisoner 'consent' when they are exploited as medical 'guinea pigs' (Chapter 3) or as work program laborers for the profit of others (Chapter 4).

Themes and content of the volume

Chapter 2: Death Row Across Species: The Execution Chamber and the Slaughterhouse focuses on the respective endgames of the U.S. Agricultural Industrial Complex and the Prison Industrial Complex: the animal slaughterhouse and prison's death row/execution chamber. I juxtapose them to show how points of shared historical geographies produced in them similar spatial designs, disciplinary technologies, and strategies of 'assembly-line' killing. The geographies of segregation, inaccessibility, distance, and concealment are essential for both. At these sites the killing itself is divided into stages, highly segregated by task and out of sight of one another, including from the workers themselves. Controlled containment and controlled mobility are integral to the functioning of the execution chamber and the slaughterhouse as well. Captives' first-hand accounts range from the accused prisoner's descriptions of the routinized anxiety, indignities, and despair endured on death row/the execution chamber, to the daily experiences of the billions of sentient animals – cows, pigs, horses, chickens, sheep, and other farm animals – herded to and through the auction block, the slaughterhouse, and other meat processing facilities that are a basic feature of today's agribusiness industry.

Putting aside the more familiar historical cases of medical and other testing on concentration camp prisoners – victims of the Nazi Holocaust or Stalin's Gulag – during much of the 20th century prisoners in 'regular' U.S. prisons were routinely used as research subjects for previously untested drugs and products. In fact, the United States was the only nation in the world to officially sanction the use of prisoners in experimental clinical trials, such as injecting them with a host of diseases from malaria to typhoid fever, tuberculosis, hepatitis, syphilis, and cholera to test their effects. *Chapter 3: The Prison as/and Laboratory: Sites of Trans-species Bio-testing* begins with an historical analysis of such medical and other testing on prisoners, examines the vast multi-billion-dollar pharmaceutical and 'health' industry behind it, and challenges practices that are today on the rise in many states despite various legal and ethical protections.

The abuse and exploitation evident in such experimentation finds clear parallels with abuse of animals as research subjects for the testing of drugs, cosmetics, vaccines, medicines, procedures, and other consumer products at laboratories located in public and private institutions and corporations throughout the U.S. Approximately 17 million captive animals are used for laboratory research and experimentation in the U.S. each year. These millions of animals, from mice to rabbits to primates, are bred as research material, and spend their entire lives in research spaces. They are typically kept in sterile concrete or wire cages; they suffer either from the isolation and boredom of captivity or the stress of crowding; and when not undergoing stressful, painful, and sometimes tortuous experiments, suffer the stress of anticipating them. Moreover, as this chapter shows, the relationship between carcerality and 'purpose breeding' extends across both animal and prisoner populations. The entire apparatus of the Prison Industrial Complex relies on an array of social, judicial, political, and economic policies to ensure criminalization, particularly of Black men who are – effectively – 'purpose bred' for prison; one in three will end up in prison at some point in their life. Thus, their vulnerability within the medical research apparatus is more closely linked to that of nonhuman animals than might appear at first glance. And as noted earlier, an important facet to my understanding of carceral spaces is how human–animal distinctions themselves are made *through* carceral processes. This chapter examines in detail relationships across racialization, criminalization, and animalization that supply the carceral logic – the vilifying logic – that governs which human and nonhuman bodies can be exploitable, disposable, and/or killable in the research lab.

In *Chapter 4: Laboring Prisoners, Laboring Animals* I examine the exploitations of labor across human and nonhuman groups and sites. I ask, what legal, political, and ethical frameworks governing property status, property relations, commodity production, and commodity relations allow prisoner and animal lives to be made exploitable and disposable? The prison as a site of captive, exploited human labor finds historical resonance with plantation slavery and subsequently with the U.S. convict lease system that lasted 50 years following the emancipation of slaves, and continues to find expression with millions of American prisoners working on the 'inside' at numerous prison industries. Prisoners make office furniture, take hotel reservations, and manufacture textiles, shoes, and clothing. Today one of the most lucrative areas for the corrections industry is in contracting out prisoner labor at tremendously deflated wages to private companies such as IBM, Boeing, Motorola, Microsoft, AT&T, Dell, Nordstrom's, Revlon, Macy's, and Pierre Cardin. And in fact one explanation for the growth in prisons over the past four or five decades is that long prison sentences maximize the availability of cheap labor, the profits of which accrue to the state or private enterprises.

Prisons are also one site where captive animals work for human benefit, and where we see the co-presence of laboring human and animal 'inmates' – horses, bulls, and calves performing with prisoners at rodeos in Oklahoma and Louisiana, or dogs working in military prisons and in regular corrections for 'therapeutic' outcomes. Laboring animals, though, can be found just about anywhere. They labor for humans at zoos, water parks, and other spaces of entertainment. The farm is probably the most ubiquitous site of animal work and labor. For millennia across the globe animals have worked in agriculture for transport, hunting, and herding. On the modern farm we find animal 'labor' producing or being produced as commodities. Animal bodies are of course transformed into meat, other foods, and clothing; but they also serve as laborers in producing other commodities – such as chickens producing eggs; pigs, horses, and cows 'contributing' extracted semen for forced (re)production of others; and dairy cows producing milk and male calves for veal production.

Juxtaposing the carceral techniques of U.S. prison labor programs with those of laboring animals allows us to see that it is particularly the linked carceral logics surrounding 'property versus personhood', instantiated with the carceral technique of animalization (and racialization/criminalization) of certain populations – and these alongside a number of linked legal potholes and mechanisms – that create the conditions under which profiteering on prison and animal bodies takes place today at a demographic scale previously unknown.

In many ways the crisis of mass incarceration, and the human rights questions posed by the ever-increasing use of solitary confinement in super-maximum prisons, maps onto the 'crisis' of the zoo and caging animals for human entertainment. In *Chapter 5: Wildspace: The Cage, the Supermax, and the Zoo*, I make comparisons across humans confined in maximum-security solitary confinement cells and nonhuman animals confined in cages in zoos and other zoo-like structures, which offer a number of opportunities to bring into parallel conversation the experience, ethics, spatiality, and politics of 'caging'. This chapter illustrates a number of overlapping oppressions and structural inequalities that span species boundaries. Caging people requires producing them as animalistic first; there is an affinity between the carceral logic of the caged animal and caged human. Further, the chapter examines the shared geographies and disciplinary regimes of the prison cage and zoo cage; the cultural and sociological 'mandates' of caging; the associated psychological-behavioral experience of being caged; and the political and ethical challenges to such long-term captivity. While the zoo and the prison are obviously different kinds of institutions, run under vastly different regimes of power, advocates for change in both the prison and the zoo have drawn on similar bio-political and ethical arguments about the cage as a disciplinary geographical space. These arguments generally place in tension 'reform' of carceral sites (dating from the Progressive Era) to abolition of them. My ongoing interest is in understanding their legal and as well 'extra-legal' regulatory regimes, as well as the productive potential of activists in reaching public, mainstream audiences with knowledge of the abuses taking place within.

In *Chapter 6: Afterword* I recap some of the intersections across human and nonhuman carceral spaces and the emergent themes from the foregoing chapters, highlighting the value of a 'trans-species' carceral geography. Carceral geographers and critical animal scholars have an opportunity to make important intellectual and policy-related linkages as they (we) advocate for progressive social transformation for both humans and nonhumans. The chapter examines the 'limits' of various prisoner and animal rights movements and the hierarchical, taxonomic, and/or speciesist carceral logics embedded within them, aligning instead with some post-humanist ways of moving forward to challenge the everyday norms of violence intrinsic to industrial production in the United States today. Ultimately the book overall speaks to both critical carceral and critical animal geographers and others who have an important role to play together in developing a trans-species ethics that is not specifically anthropocentric.

2 Death row across species

The execution chamber and the slaughterhouse

Introduction

It is hard to miss the parallels in the rise and development of the Agricultural Industrial Complex (AIC) and that of the Prison Industrial Complex (PIC) in the United States in the last half century. Both of these industries rapidly expanded during the 1970s and 1980s era of deregulated 'big agriculture' on the one side, and changes in drug and sentencing laws that led to new thresholds in mass incarceration on the other – with an unprecedented 2.4 million people behind bars in the peak year of 2014.[1] In this chapter I hurtle past many social, political, and economic issues that could present themselves for analysis, and proceed straight on to discussing what must be considered the crucial endgames of these respective industries: I focus on these as sites of killing and death in order to examine and enlist in conversation their shared carceral logics, spaces, operations, and technologies. As Thomas and Shields (2012: 4) observe, "state-sanctioned killings are capital punishment in one arena and 'processing' in another"; they intersect with a concern for "human and animal's [sic] lived experiences."

In line with these authors' sensibilities I begin the chapter with a brief overview of capital punishment in the U.S., highlighting the embodied experiences of prisoners living decades on death row, of inhabiting the life and death of the condemned. I place these experiences alongside those of livestock animals subjected to the violent assembly-line killing of the slaughterhouse. The practices and forms of confinement within the prison's 'death house' and the livestock slaughterhouse are notably similar in their functional/physical form and modes of operation. My discussion highlights the ways that the prison death row/ execution chamber and the animal slaughterhouse have been similarly designed to kill and dispose of bodies, even if the intentions behind them are obviously different – killing to punish or simply eliminate, versus killing for food or other commodities. At these sites, though, we see similar spatial designs; similar technologies of surveillance, regulation, and movement of bodies through them; and similarly choreographed and segregated activities of assembly-line type killing.

My discussion rests primarily on conceptual categories of these carceral spaces rather than on any specific sites or institutions; offering a critical analysis of the shared carceral logics that inform their killing operations. Throughout the chapter I wrestle with key questions surrounding the epistemic violence characteristic of these institutions. What makes particular mechanisms of violence useful in these particular carceral domains? What is it that makes their characteristic functions, technologies, and practices attractive to their respective regimes of power? One significant element I bring forward is their respective 'politics of sight' (after Pachirat 2011; Elder et al. 1998); that is, the role of proximity, distance, concealment, and visibility in their exercise of power, domination, and death.

At the outset and as an entrée to the chapter generally, I note that there is plenty of empirical evidence and reason to define the death chamber and slaughterhouse as respective "carceral spaces" due to their *detrimental* carceral effects of *confinement* (after Moran et al. 2017, see Chapter 1). Captives' first-hand accounts of their experiences – from the accused prisoner's description of the fear and despair of knowing the approach of death to the bleating and shrieking of animals prodded, stunned, and whipped to unknown futures in the slaughterhouse – all provide valuable empirical evidence of the emotional, psychological, and physiological strain and trauma of inhabiting the body of 'the killable'. Following on the important methodological points raised in Chapter 1, though, it is important to keep in mind King's (2013: 6–8) admonition that while animals feel sorrow, "goat grief … is not chicken grief. And chicken grief is not chimpanzee grief or elephant grief or human grief. The differences matter." In other words, it is not my intention to somehow prove that these sites of assembly-line killing produce 'the same' embodied experiences or responses in their human and nonhuman sufferers, or that any one type of grief is more consequential or meaningful than another.

Rather, King's is an important insight because the problematical human–animal divide can be reinforced when we say that forms of pain and suffering are the same or similar. Wise (2000, 2002), for example, has argued for the similarities across humans and nonhumans as a way to extend compassion and empathy to nonhuman species and grant them 'personhood'. And this has, in some ways, destabilized the human–animal boundary, encouraging us to think that some animals are not so different from humans (see detailed discussion in Chapter 3). And yet, such does not dissolve the human–animal boundary – it just shifts it, and more problematically, encourages political and legal protections only for those nonhuman animals who are recognizably 'like us'. Meanwhile, those who are *not* like us – such as billions of livestock animals herded to and through the slaughterhouse – whose emotional capacities or cognitive capabilities are not as recognizable, remain 'animalized', and become even more vulnerable to violence, exploitation, and killability.

The human population most at risk for suffering on death row in the first place – predominantly Black and other minority men – have been thoroughly animalized through their incarceration, racialization, and poverty. Yet as I also discuss in the next chapter, pointing out the many ways in which death row or other prisoners have been cast as 'subhuman' or 'animal' does not mean that I equate their experiences to that of animals (e.g. Hart 2014: 673–674; Coetzee 1999; Kim 2011: 313, 326–330). But it does suggest that the animalization of certain humans and certain nonhumans within these carceral settings works toward similar ends. 'Animalization' as a process subjugates and makes disposable both humans and nonhumans and is rooted in deeply uneven hierarchies of power that provides certain groups with rights and resources and denies them to others (Lopez and Gillespie 2015: 3–7). It is what provides the carceral logic behind capital punishment as well as the killing of billions of animals for food and other consumer products each year. Thus, at the outset I simply would acknowledge the alterity of species' experiences, minds, and emotions, but insist nonetheless that human and nonhuman experiences within the death house or slaughterhouse are profoundly impacted in significant ways via entangled carceral logics, structures, and operations.

State-sanctioned killing: the execution chamber

Do you understand what will happen to you?
Do you have any questions?
What do you want for your last meal?
Do you plan to make a last statement?
What do you want us to do with your body?
What do you want to do with your property?
Who do you want to have your money?
Who will witness your execution?
Do you know what we expect you to do?
Are you comfortable?
If not, what can we do?
If your stay is denied, who do you want to call?
What color clothes do you want to die in?

> Questions asked of death row prisoners in preparation
> for execution (Malon with Hawkins 2007: 131–134).

The parallel rise in mass incarceration and retention and expansion of the death penalty are indicative of an historical "hearty American appetite for punitiveness" that is exceptional among Western nations (Steiker and Steiker 2014; Garland 2005; Alexander 2012; Tyner and Colucci 2015). Today there are 36 prisons in the U.S. with a "death row" – two federal prisons (Terre Haute, Indiana and Fort Leavenworth, Kansas), and 34 state prisons. At the latest count these prisons hold 2,905 people on death row awaiting execution (Death Penalty Information Center 2017) – 42 percent of whom are Black, 42 percent White, 13 percent Hispanic, and 3 percent 'other'. Clearly these percentages indicate the uneven and racialized application of the capital sentence considering that, for example, Black people comprise 34.4 percent of those executed in the U.S. since 1976, and 42 percent of those on death row, yet comprise only 13.2 percent of the

U.S. population (Death Penalty Information Center 2017). Most death row prisoners await execution by lethal injection, notwithstanding numerous recent 'botched' executions, drug shortages, and states seeking alternatives, such as Utah lawmakers seeking to reinstate the firing squad (e.g. Hannon 2015).

After the temporary abolition of capital punishment in 1972 (in *Furman v. Georgia*), the death penalty 'came back with a vengeance' in the years following its reinstatement by the U.S. Supreme Court in 1976 (in *Gregg v. Georgia*; see Gottschalk 2006; Steiker and Steiker 2014). By the late 1990s, death sentencing rates and execution rates reached highs that the U.S. had not seen in 50 years, with the Supreme Court leaving the legalities of capital punishment up to individual states. Today, state sponsored execution is legal in 32 states, while 18 have abolished it. (Nebraska abolished it in 2016 but that same year reinstated it with a politically-charged referendum vote; Bosman 2015a; 2015b.[2])

Since 1976 there have been 1,442 state-sponsored executions, with 34.4 percent of those Black, 55.8 percent White, 8.3 percent Hispanic, and 1.6 percent 'other.' Of these, 1,267 executions have been by lethal injection, 158 by electrocution, 11 in gas chambers, three hangings, and three deaths by firing squad. Over 75 percent of the murder victims in cases resulting in an execution were White, even though nationally only 50 percent of murder victims are White (Death Penalty Information Center 2017; *Death Row USA* 2015). Christianson (2010), Gillespie (2003), Gottschalk (2006), and the Death Penalty Information Center (2017) offer insights into these and other historical trends – one of which is that today, for the first time in half a century, death sentences, executions, and public support for capital punishment has declined – although whether this is a trend or blip in the trends is uncertain, particularly as 'populists' continue to support it (Toobin 2016).[3]

Prisoners sentenced to death typically spend over a decade awaiting execution, some for over 20 years or more due to lengthy appeals processes. They spend 23 or 24 hours a day alone in their cells, living with the constant anxiety of impending death. As the execution date approaches they spend up to two weeks in a separate 'death watch' cell near the execution chamber itself. Cunningham and Vigen (2002) provide a useful overview of the psychological and other mental and social disorders among prisoners sentenced to death. Not surprisingly, they suffer high rates of mental illness, what some have termed the 'death row syndrome', which would of course be in addition to the mental illnesses caused by solitary confinement in and of itself (Guenther 2013; Haney 2008; Rhodes 2009; see Chapter 5). The psychologically torturous conditions of death row also can lead to state-produced 'death row volunteers' among those who cannot withstand the cruel and torturous conditions and to whom death is preferable (Mulvey-Roberts 2007; Rodriguez 2005).

Much debate surrounds the practice of capital punishment, including those about its underlying racial disparities, controversial methods of killing, inconsistent and arbitrary patterns across state lines (with the South disproportionately represented), its (lack of) efficacy in deterring crime, its cost, and of course its basic (im)morality (Cunningham and Vigen 2002; Equal Justice USA 2016).

The issue of secrecy surrounding the execution chamber is another, with only a tiny number of court-approved spectators allowed to witness the killings. Nevertheless, 22 electric chair executions in Georgia in the 1990s were recorded by members of the state's Department of Corrections and later released via Public Radio as *The Execution Tapes* (1998). *The Tapes* featured, among other things, gruesome narrations of executions that had to be "reinitiated" when prisoners failed to die after two minutes of electrocution. One portion featured the last words of Jerome Bowden, who, with an IQ of 59, betrayed with his last words a probable lack of grasp of what was happening to him:

> I would like to thank the people of this institution for taking such good care of me in the way that they have. And I hope that by my execution being carried that it may bring some light to this thing that is wrong … Thank you very much.[4]

During the same period National Public Radio also released *Witness to an Execution* (2000), which tells the stories of the corrections personnel involved with 100 executions by lethal injection in Huntsville, Texas. *Witness* offers a minute-by-minute description of events – the last day visits, the 'walk' to the execution chamber, last phone call, last meals, last words, and the 'go ahead' signal of the warden. According to those assisting officers: "Some of them are crying … Some of them have been sweating. Some of them will have the smell of anxiety, if you will. Of fear" (ibid.; excerpted in Mulvey-Roberts 2007: 145–154).

A number of 'death row diaries' offer first-person accounts of the torturous experience of living on death row and the days leading up to execution (*Death Row Diaries* 1999; Basu 2013; Mulvey-Roberts 2007). These diaries attest to the emotional and psychological fear, despair, and terror prisoners experience, as well as document state-sanctioned torture, brutality, and violence. In 2003 Billy Frank "Sonny" Vickers wrote about his impending execution at the Huntsville, Texas death house. By midnight his death warrant had expired and he was placed back on death row, although he was eventually put to death a month and a half later. Dow et al. (2004) describe his case as one illustrating the 'banality of death' of an (uncontested) guilty prisoner whose constitutional rights were nonetheless ignored. Vickers (2014) offered a valuable description of such an experience:

> My thoughts are broken when the warden comes into the death house to tell me what will be taking place … If I can't walk, they will carry me, but either way I'm going.
>   The chaplain comes and tells me, while I'm on the gurney he will be there holding my ankle to offer comfort. I see the door that the ambulance will back up to, to pick up my body and that's when it strikes me all over again, "this is it." There's no way to describe the pressure I feel as I pray they'll hurry up and get it over with.

Every time the walkie-talkie bursts to life, a door slams, the phone rings, I nearly jump out of my skin. This is almost constant for six hours. The chaplain tells me that if I hear rustling and movement in the back, he says it's just the execution team getting ready and for me not to be "alarmed" (they're just coming to kill you. Don't be "alarmed"!).

I've been broke out in a cold sweat for 2 hours. Can't think. Just pace, pace, pace. Back and forth, back and forth, 3½ steps … I eat some of my last meal but I can't taste a thing. I just look down and see that some of it is gone.

Six o'clock comes. Nothing. Seven o'clock. 8 o'clock. Same thing. My mouth is so dry no amount of water can wet it. I know they're going to open that door any minute and confront me with that gurney and those needles. This is it. This is it. Every time I blink the sweat out of my eye I see it open, I think, that door.

William Van Poyck wrote to his sister about his experience of 40 years in prison, 26 of them on death row at Florida State Prison (Solitary Watch 2013). Excerpts include:

February 9, 2012: Yesterday the prison was locked down all day for the standard "mock execution," the practice run which occurs a week prior to the actual premeditated killing … to make sure the death machine is in working order, everyone on their toes.

February 25, 2012: Robert Waterhouse was scheduled for execution at 6:00pm this evening [but] he received a 45-minute stay which morphed into an almost 3-hour endurance test.

[P]ause for a moment to imagine being on that gurney for over three hours, the needles in your arms. You've already come to terms with your imminent death, you are reconciled with the reality that this is it, this is how you will die, that there will be no reprieve. Then, at the last moment, a cruel trick, you're given that slim hope … You are totally alone, surrounded by grim-faced men who are determined to kill you. Your heart pounds, your body feels electrified and every second seems like an eternity as a Kaleidoscope of wild thoughts crash around franticly in your compressed mind. After 3 hours you are drained, exhausted, terrorized, and then the phone on the wall rings and you're told it's time to die.

May 22, 2013: Today my neighbor, Elmer, went on Phase II of death watch, which begins 7 days prior to execution. They remove all your property from your cell while an officer sits in front of your cell 24/7 recording everything you do.

May 3, 2013: Today Governor Scott signed my death warrant and my execution date has been scheduled for June 12th, at 6pm. When your warrant gets signed so many things suddenly become trivial. My magazines and newspapers stack up unread; I have little appetite to waste valuable, irreplaceable hours reading up on current events. The other day I caught myself reaching for my daily vitamin. Really? I wondered, as the absurdity hit me. Likewise, after 40 years of working out religiously … what's the point?

In Van Poyck's last entry, dated 15 days before he was executed, he wrote:

> *June 12, 2013*: I read in a recent newspaper article that the brother and sister of Fred Griffis, the victim in my case, are angry that I'm still alive and eager for my execution. I have thought of Fred many times over the years and grieved over his senseless death.

★ ★ ★

As the above discussion highlights, a great deal of ritual accompanies the killing of a human life through capital punishment. The rituals surrounding the last meal, clothes selection, the chaplain visit, the staff practice-run in weeks leading up to the execution, the selection of viewing audiences and their placement in the viewing gallery – not to mention the typically lengthy appeals process and stays of execution that can last until the final hour or minute – all signal a highly ritualized form of killing that reflects the seriousness accompanying the taking of a human life, even a so-called criminal human life.[5] While we might note a great deal of 'ritual' accompanying some animal killings, such as within religious ceremonies (Robbins 1998; Elder et al. 1998; Casal 2003; Miele 2013), the industrialized slaughter to which I turn next appears wholly absent of such ritual and ceremony. This signals, among other things, the relatively high value placed on human life compared with that of animal life. Moreover, the vast scalar difference between animal slaughter (below) and the killing of death row prisoners are worth pointing out: 1,442+ prisoners have been executed in correctional institutions over a 40-year period, versus 10 billion land animals each year. The magnitude of this difference, again, signals much about the relative killability and grievability of human and nonhuman life generally, even if some lives within these taxonomic categories are clearly imputed with greater or lesser value (Butler 2009; Deckha 2010; Glick 2013; Cacho 2014; Kim 2015).

Entangled experiences: the slaughterhouse

Historical geographies of the animal slaughterhouse are instructive reminders of the origin of the 19th-century cattle towns that fed Chicago's Union Stockyards in the 1850s and 1860s, setting the stage for today's meatpacking industry (Patterson 2002; Coetzee 1999; Rifkin 1992; Philo and MacLachlan 2017). Patterson (2002: 57–58) describes the enormous complex of hotels, restaurants, saloons, offices, and "an interlocking system of 2,300 connected livestock pens" that took up more than a square mile in southwestern Chicago. At the time, the meat companies Armour and Swift each employed more than 5,000 workers within those yards. By 1886, more than 100 miles of railroad track surrounded the yards, and each day trains with new refrigeration capability unloaded hundreds of cars full of western longhorn cattle, sheep, and pigs. This first 'mass production industry' introduced the conveyor belts, suspension hooks, scraping and skinning machines, and other technologies to increase speed and efficiency (Rifkin 1992; Gideon 1948: 213–246), and by 1900, 400 million livestock were slaughtered annually. Today, U.S. slaughterhouses kill that number of animals in less than two weeks (Patterson 2002: 58–64; Pachirat 2011).

As in the prison, the day-to-day embodied experience of captivity; of being identified with a number, a tattoo, a brand, and other forms of bodily modifications; the strain of knowing the approach of death or of the stunning apparatus or whip; all are interwoven into the day-to-day carceral space of the animal slaughterhouse (Gillespie and Collard 2015). Today, approximately 10 billion land animals are held captive, mutilated, and killed in the U.S. meat-producing industry each year (9 billion of them chickens), and this figure does not include the billions more sea creatures who are counted not per animal but by weight (by the ton). These billions of cows, pigs, horses, chickens, sheep, and other farm animals herded to and through the auction block, the slaughterhouse, and other processing facilities are a basic feature of today's agribusiness industry. Industrialized food production is where, by far, the most violence against animals occurs.

Gillespie (2014), Collard and Gillespie (2015), Urbanik (2012) and many others have offered empirical evidence of the sentient experience of animal life being made into meat. Gillespie (2014) describes the everyday humiliations, suffering, indignities, and sadnesses of cows, bulls, and calves being separated from family and produced as 'gendered commodities' in the dairy and meat industry. Collard and Gillespie (2015: 6) describe the "bellows, shrieks, bleats, grunts, and squawks" of animals arriving at the auction yard in transport trailers, who are unloaded and herded into a series of chutes and holding pens in preparation for assessment, sale, and other doomed futures.

With the slaughter of cattle and horses, a 'captive bolt stun gun' is used to shoot a five-inch metal bolt to penetrate an animal's brain at the start of the assembly line kill. This attempt to render an animal insensible to pain by means of a rapid single blow before being processed was a result of the 1958 Humane Methods of Livestock Slaughter Act, ostensibly intended to make the slaughter of farm animals "more humane." (It is the only federal law that 'protects' farm animals; Higgin et. al. 2011; Wolfson and Sullivan 2004: 205–208.) Not only does the stun gun procedure not prevent cruelty, but the procedure often does not work. Pachirat (2011) describes the entire assembly line kill in ghastly detail from his experience working undercover at an Omaha slaughterhouse. He describes how 'the knocker' hydraulically controls the movement of the conveyor that holds and constricts the animal's body, and which presses its head out of a rectangular box for the stun gun:

As the bolt retracts, gray brain matter often flies out of the hole in the cow's skull, sometimes splattering the clothing, arms, or face of the knocker. Seconds later, blood gushes out of the wound … whether the head falls or the neck stiffens, the cow's eyes typically take on a glazed look, and its tongue often hangs limply from its mouth. Sometimes the power, angle, or location of the steel bolt shot is insufficient to render the cow unconscious, and it will bleed profusely and thrash about wildly while the knocker tries to shoot it again (2011: 54–55; also see Schlosser 2001: 172–174).

At this point the cow is dropped onto a conveyor, conscious or not, and hooked up by the left hind leg, often kicking wildly. And meanwhile, of course, "the next cow watches everything. Then her turn comes. This is Dante's Inferno" (Patterson 2002: 68; see Fig. 2.1).

As with cattle, in the slaughter of horses a stun gun is used to shoot a metal bolt to penetrate the animal's brain from the front, a process brought gruesomely to light in George Franju's historic film of a Paris slaughterhouse, *Blood of the Beasts* (1949). American horses are currently transported long distances for slaughter in Canada and Mexico, but proponents are currently working to reopen U.S. plants (Horse Slaughter 2015). Pollan (2002) describes the contemporary hog operation. Among other atrocities, due to premature weaning, confined pigs chew on the tails of those caged in front of them to the point of infection. The industry response is typically 'tail docking', crippling the animal's tail with pliers to produce a more sensitized stump so that the "demoralized pig will mount a struggle to avoid" being chewed (also see Wise 2009).

Figure 2.1 A cow awaits the stun bolt gun. Photograph courtesy of Slaughter – Occupy for Animals (France 2007).

By the numbers, chickens are the most abused animals on the planet, with more than 9 billion slaughtered each year in the U.S. alone. Striffler (2005) examines the transformation of the chicken industry from the family farm to industrialized slaughter, focusing on the incredible volume of chicken consumption originating in President Herbert Hoover's 1928 promise of a "chicken in every pot," the industry's impact on (mostly immigrant) workers and their working conditions, and the unhealthy consumption habits of the 'McNugget' generation (Striffler 2005: 7; Schlosser 2001). Chickens and other poultry are not covered by the Humane Slaughter Act (Wolfson and Sullivan 2004). Most spend their entire lives, from hatching to slaughter, in total intense crowded confinement. Pollan (2002) argues that egg-producing hens by far 'have it the worst', although chickens, turkeys, and other poultry across the spectrum exist in severely overcrowded small cages leading frustrating and boring lives, unable to do anything but stand and eat, unable to scratch, clean themselves, build nests, or spread their wings or even turn around (see Fig. 4.4). At six or seven weeks old they are crammed into cages and delivered to slaughter. Live birds are shackled upside down on an automated line where they are electrically stunned, passed through an automatic device that cuts the jugular vein, and then thrown into scalding water, oftentimes while still conscious, to remove their feathers. And as Jones (2014: 94) reports, 'broiler chickens' who have been liberated from farming situations "still shriek after several months of living at a sanctuary" when they are lifted into coops by humans.

Space, technology, and control

Numerous examples of the symbolic relationship between the violence of prison torture and execution and the animal slaughterhouse can be found, such as Brower (2004: 1360–1362) illustrates via images from the notorious Abu Ghraib prison in Iraq where prisoners were physically handled and photographed as animals to be slaughtered. Guards used blades to cut away prisoners' jumpsuits, from their necks to their thighs, branding prisoners like cattle, drawing words and symbols on their legs and buttocks, and forcing them to crawl like dogs on their hands and knees, to bark on command, and to follow their captors on leashes or strings. The play *La Historia de Nuestras Vidas* (*The Story of Our Lives*) describes the personal experiences of 400 undocumented workers who were arrested in the 2008 raid of the AgriProcessors Meatpacking Plant in Postville, Iowa (Skitolsky 2008; Raymond 2009; see Chapter 4). The play describes the workers' experiences of being held in a former site of cattle slaughter, and their newly raised consciousness about the treatment of animals at the site.

These symbolic referents provide powerful insights into the relationships of violence across the Prison Industrial Complex and the Agricultural Industrial Complex. They provide a useful springboard to the material geographies of these sites and the ways that systematic violence at them is carried out; through their shared structural designs and disciplinary technologies and practices they terrorize animal and human bodies in similar ways. The material geographies of the prison execution chamber

and the animal slaughterhouse especially map uncannily well onto one another –
their locations; their physical structures, spatial layout and design; as well as their
technological and other control features that regulate movements within them.

The power and politics of sight

Most obviously, these carceral sites are 'hidden in plain view' in rural or remote
locations, their architectures so innocuous and ordinary that they do not attract
attention (Fig. 2.2). From an aerial view, as Merritt and Hurley (2014) argued,
the prison and slaughterhouse look the same:

> Central to these industries is the invisibility of their operations. Both of
> them rely on a "post-regulatory" systematized objectification of bodies, the
> visceral nature of which would be publicly inflammatory, and thus detri-
> mental to economic profit. Therefore, these industries have created hidden
> geographies that conceal their physical locations and processes while at the
> same time normalizing the notion that Americans need prisons to stay safe
> and meat to stay healthy.

Figure 2.2 Aerial view of the banal landscape of the Federal Correctional Institution
FCI-Herlong, California. M. Spaulding, *Federal Correctional Institution, Men-
dota, California: Project Description and Employment and Business Opportunities*
(Washington, D.C.: U.S. Department of Justice, Federal Bureau of Prisons,
2005).

Pachirat (2011: 23) discusses the "banal insidiousness" of the slaughterhouse that hides in plain sight, its construction blending physically into the landscape of 'Everyplace USA':

> This city block-wide windowless box of gray corrugated steel set on a shoulder-high slab of concrete, topped by massive ceiling fans, surrounded by black asphalt, a chain-link fence, and guard huts, and fronted by a modern office complex of glass and aluminum presents only the generic face of mass production. The building materials, size, structure, and layout of the slaughterhouse could pass for the community college to the south, the tool factory to the east, or the pet-supply store to the north. Only the coming and going of hole-pocked semi-trailers, the rhythmic clanging of hoofs, and the omnipresent stench intimate what lies inside.

As Glick (2013: 645) observes, such places are 'everywhere and nowhere'. "This dynamic must be understood as the very definition of necropolitical regimes of terror, in which the spaces and moments of 'everyday life' are marked by the sign of constant death."

Sites of capital punishment today also inhabit a set of insidious visual banalities, at least in comparison to the past. The last U.S. public execution occurred in 1936. Once ritual executions were moved from the public square, the public spectacle of death shifted to the sterile courtroom, where the suffering of victims and judicial process became the important ceremonial stage of punishment (Gottschalk 2006: 199–203; Lynch 2000). In this way the infliction of punishment shifted to the imposition of the death sentence in court proceedings. Thus the loud, unruly, festive spectacles of public execution were replaced by executions carried out in the private space of the jail or prison yard. In this respect prisons must be understood as categorically developing into places not of reform and rehabilitation, as the Bureau of Prisons mandated them when it was established in 1930, but rather as places of violence and killing (Morin 2013).

Gottschalk (2006: 199–216) provides a useful analysis of the origins and development of the carceral state, specifically as it relates to the sovereign state's right to take life via capital punishment. To Gottschalk, the death penalty has been framed in particular kinds of ways, which is to say that the 'passions of the public' – particularly those rallying around victims' rights – became a central and legitimizing force in shaping of the state's penal policy. Such punitive tendencies helped "lock in the carceral state," with the effect of magnifying the suffering of all victims and conversely, the immorality and evilness of all offenders (Gottschalk 2006: 218–219). Moreover, an 'informed public' has not, by and large, repudiated the horrors of capital punishment when confronted with its realities. Instead, public confidence in the carceral state has been bolstered by 'law and order' politics and politicians (Steiker and Steiker 2014: 231). And again, it is within the site of the courtroom and judicial sphere (not the legislative) where the power of the carceral state, via the politics of sight – particularly as articulated and demonstrated by the survivors of homicide victims – has been most effectively codified.

At a more generalizable level Elder et al. (1998: 78–79) assert that it is the site of harm that is the most crucial aspect in determining the ostensible 'legitimacy' of violence – and, I would add, the legitimacy of carceral space itself. They observe that all societies have culturally acceptable sites and contexts for torture, killing, and abuse, and that the same act or practice could be considered savage in one context while perfectly acceptable in another. All sorts of cruel acts are condoned when out of sight, yet beyond those contexts – e.g. killing a cat or other animal on a U.S. street corner – the same acts can even be criminalized (Elder et al. 1998: 85; Schlossberg 2014; Cacho 2014). Thus as Emel and Urbanik (2010: 208) argue with respect to factory farms (but which could also be said of the execution chamber and pharmaceutical testing lab), geographers have made important contributions to the study of the ethics of such spaces precisely by paying attention to the banal invisibility of their locations. When the torture and violence occurring at such sites are made visible, there is public outcry, at least for (some) animals. The legitimacy of the killing also revolves around who is doing the killing – what 'others' and their practices have legitimacy – as well as who is being killed (Deckha 2013b; Kim 2011; 2015). Which animals and humans do we care about, and which not? Deeply embedded in carceral logics is how human–animal distinctions themselves are made, and how then hierarchies of 'worthiness' and value or disposability and killability of certain humans and nonhumans are made possible.

Pachirat (2011: 3–14) helpfully analyzes the politics of visibility by arguing that geographies of segregation, inaccessibility, distance, and concealment are mechanisms basic to the exercise of power; they are essential for the meat-packing industry as well as the prison/execution chamber and other zones of confinement (also see Deckha 2010). Thus making something visible or invisible itself is a mechanism of power. Power might work by demolishing distance (e.g. via global satellite systems), by making visible what is concealed in order to control it; but it also works through the creation of distance, isolation, and concealment of what is physically and morally repugnant (Pachirat 2011: 233–256). Transport of animals to slaughter, to take another example, tends to be performed at night and in unmarked trailers and trucks[6] – not unlike the generic, unmarked vans used to transport prisoners.

Spatial organization of assembly-line killing

At carceral sites and institutions, the killing itself is divided into stages, highly segregated by task and out of sight of one another, including from the workers themselves. The police shooting of a cow who escaped from a meatpacking plant to the streets of Omaha, Nebraska caused public outrage as well as outrage by the plant's workers – suggesting that these workers failed to connect the shooting of this individual to the slaughter of one of them "every twelve seconds" at their workplace (500,000 every year at that site alone; Pachirat 2011: 5–15).

Their work requires that the spatial organization of the site be compartmentalized such that the violence is so thoroughly routine as to be 'invisible' to the workers themselves, but also requires a deep compartmentalization and desensitization to the nature of their workplace.

Pachirat (2011: 44–59; see Fig. 2.3) provides a number of provocative maps that illustrate the geographies of the slaughterhouse floor that, for example, separate 'life' from 'death' – albeit with recognition that sentient beings can be messily and clearly alive and conscious long past applicable boundaries. He describes the kill floor stages and the points at which the cow 'loses its tail, its hoofs, its hide, its heart, its head, its lung, liver, intestines'. They die ... piece by piece" (Schlosser 2001; Pachirat 2011: 53–67). At the Omaha slaughterhouse there are 121 job functions – "121 experiences of industrialized killing" – alerting us to what each worker would be able – or more likely, unable – to see of the entire operation from any particular workstation. Most workers operate in the 'zone of death' – only able to see already dead, disembodied, decontextualized pieces of carcass. Thus another way that invisibility works is to keep workers themselves separated, not only on the slaughterhouse floor but throughout the physical plant. Those on the 'dirty' versus 'clean' sides of the operation must use separate lockers, toilets, showers, bathrooms, and lunchroom, ensuring they will not come in contact with one another or experience, together, animals being transformed from creature to carcass.

Patterson (2002: 110–131) argues that industrialized "killing centers" have several things in common: their technologies, speed, efficiency, and 'rational' scientific management and Tayloresque assembly-line techniques (also see Glick 2013; Higgin et al. 2011: 175; Schlosser 2001: 172; Giedion 1948). Humans have been confining and killing each other and animals for millennia, but the specialization and mass-production characteristic of the modern industrial era – the "mechanization of death" – was something new and perfected by the late 19th century (Giedion 1948: 240–214). Within these spaces are routine, mechanical, predictable, repetitive, and programmed practices. "Just the right mix of deception, intimidation, physical force, and speed is needed to minimize the chance of panic or resistance that will disrupt the process" (Patterson 2002: 110). Controlled containment and controlled mobility are integral to the functioning of the slaughterhouse, as again, they are equipped with an array of chutes, pens, ramps and technological equipment intended to efficiently and quickly move animals for processing; in addition to the stun gun described above are mobile shackle lines, electric prods, hoists, and mechanical restraining pens (Gillespie 2014; Higgin et al. 2011). Purely with respect to spatial organization, mass killing requires routine, mechanical, repetitive, and 'programmed' procedures and tasks.

* * *

Numerous news stories abounded in April 2017 about the 'execution assembly line' in Arkansas as the state was preparing to execute eight men in an 11-day

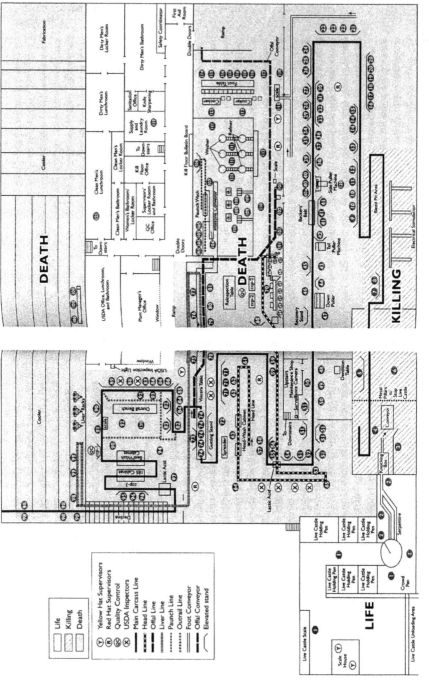

Figure 2.3 Divisions of life, killing, death at an Omaha slaughterhouse. From T. Pachirat, *Every Twelve Seconds: Industrialized Slaughter and the Politics of Sight* (2011: 58–59). Reproduced courtesy of Yale University Press.

period before the expiration of its supply of the drug midazolam, one of the drugs used in its lethal injection protocol. Headlines about what Cobb (2017a) described as a "banal horror" shouted across the media spectrum: "Arkansas Rush to Execute 8 People"; "Arkansas Is Turning Death Penalty into an Assembly Line"; "Arkansas Inmate Speaks Out on Assembly Line"; "Mentally Ill Man Set to Die in Arkansas Conveyor Belt"; and "Time Running Out to Halt Arkansas Execution Assembly Line." The state's killing spree ended with four executions in rapid succession, including two in one day, three hours apart. (Killed were Ledel Lee, Jack Jones, Marcel Williams, and Kenneth Williams.) This assembly-line metaphor for what was happening in Arkansas was perhaps a more apt description of the capital punishment process than the writers of these headlines might have envisioned.

In the prison's death row and execution chamber, each movement is carefully choreographed, regulated, and surveilled. Lynch (2000: 15) and Gillespie (2003) describe the "execution assembly line" of the prison's death house and execution chamber. Prisoners remain in their regular death row cell until two weeks prior to execution, then are moved to a specially designed 'death watch' cell close to the execution chamber. Here the minute details of the condemned's final weeks in isolation are tracked; closed circuit TV and computer tracking systems open cell doors, control lights, and maintain an activity log of every movement and location of the prisoner within the death house cell. No conversation with guards or other prisoners is allowed, and inside the cell the prisoner is allowed only the most basic amenities: toothbrush, toothpaste, small bar of soap, washcloth, and towel – no TV or radio. As Mulvey-Roberts (2007: 126–130) and Lynch (2000: 15–16) argue, there is no reason to isolate a prisoner in this manner for two very long weeks except perhaps to force him to withdraw inside himself. The presence of a diaper closet, which will on the appointed day make the "waste disposal process" easier for those who are cleaning up, must weigh heavily on the prisoner's psychology (Lynch 2000: 17). Hemsworth (2015) adds an important intervention into this 'regime of terror', alerting us of how important sound becomes in the pre-execution unit. In fact, the affective experience of the sonic itself may be harder to ignore than the visual, with every sound, including the most mundane, producing terror as it signals a preparation for killing and death (cf. Vickers 2014 above; also see Chapter 5 and Brower 2004; Gregory 2006; Guenther 2013; Haney 2008; and Morin 2013 for more general discussion of physical and mental torture within the prison setting).

In the execution chamber itself, specific tasks are delegated to each member of the execution team, including the final enunciation – 'go ahead' – of the warden (Lynch 2000: 16). As noted by a guard assisting with a lethal injection at Huntsville, Texas (above), "Usually within about twenty seconds he's completely strapped down. Twenty to thirty seconds. I mean, it's down to a fine art" (*Witness to an Execution* 2000). Transcribed narratives of the warden and his staff include the likes of:

Dean: I've participated in approximately over a hundred executions as a member of the tie down team. Each supervisor is assigned a different portion – like we have a head person, a right arm, left arm, right leg, left leg. And the right leg man will tell him "I need you to hop up onto the gurney. Lay your head on this end, put your feet on this end." Simultaneously while he's laying down the straps are being put across him.

Green: I'm a member of the tie down team in the execution process. What I do, I will strap the offender's left wrist. And then there are two belts – one that comes across the top of his left shoulder – and then another goes right straight across his abdominal area.

Brazzil: After they are strapped down then all the officers will leave. And then it's the warden and myself in the chamber with him, and there'll be a medical team come in and they will establish an IV into each arm.

Brazzil: I usually put my hand on their leg right below their knee, you know, and I usually give 'em a squeeze, let 'em know I'm right there. You can feel the trembling, the fear that's there, the anxiety that's there. You can feel the heart surging, you know. You can see it pounding through their shirt.

Moritz: The warden will remove his glasses, which is the signal to the executioners behind a mirrored glass window. And when the glasses come off, the lethal injection begins to flow.

Brazzil: After they're strapped down and the needles are flowing and you've got probably forty-five seconds where you and he are together for the last time. The conversations that took place in there were, well, basically inde-scribable. One of them would say "What do I say when I see God?" You've got forty-five seconds and you're trying to tell the guy what to say to God?

Gideon: You'll never hear another sound like a mother wailing when-ever she is watching her son be executed. There's no other sound like it. It is just this horrendous wail. You can't get away from it. That wail surrounds the room.

Green: One thing I am glad of is that we're not using the electric chair. I don't think I would want to be part of that. This process here, it's clinical. The inmate, other than the fact that he's expired, you don't know anything has happened to him. And, you know, that's good.

At some precise moment during this process (e.g. '6:09'), the warden directs the staff to escort witnesses into the small rooms adjacent to the death chamber and they "push up real close to the windows to get a view." Lynch (2000: 19), who served as a witness to the execution by lethal injection of "Inmate # 85271," Arthur Ross, in 1998 in Arizona, describes the intricately choreographed process by which hierarchically organized groups of witnesses enter and leave the execution viewing area. When the curtains are drawn: "the prisoner is already strapped in on gurney ... lying immobile in a brightly lit, white tiled, sterile looking room. He is covered by a sheet to conceal the injection sites ... The Warden enters the room, turns on the microphone and reads the death warrant."

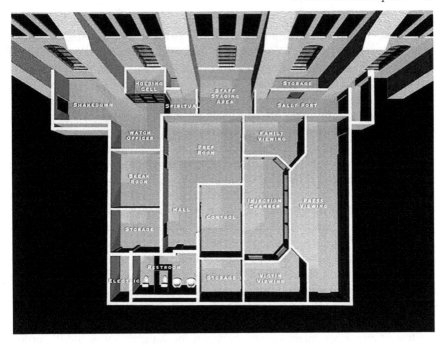

Figure 2.4 CAD model of San Quentin's lethal injection facility. California Department of Rehabilitation and Corrections, n.d.

Sperry (2014: 3–4) examines the layout and architecture of the San Quentin lethal injection chamber, describing the banal-looking "injection room" as akin to a dentist's office in a strip mall. The design of facility overall (Fig. 2.4) provided, among other things, better viewing areas for witnesses and "more workspace around the body of the condemned man" than their former gas chamber room. To Sperry, the chamber "teeters on the edge of visibility and invisibility" since the floor plan and design were made public in order to prove their 'constitutionality'. The position of the lethal injection team:

> is tightly prescribed .... regulations detail which of the various execution subteams will occupy different rooms, and when individual members will enter or leave them. A small, one-way-mirrored window and four hose ports connect the intravenous subteam, who insert the IV lines into the condemned man after the security subteam has brought him into the injection room and strapped him down, and the infusion subteam, who push the lethal chemicals through the syringes in the Infusion Control Room behind the condemned man's head.

Summarizing the scene, Sperry (2014: 4) writes: the "banality of California's execution chamber is just as troubling as if it were a Gothic Revival chamber of horrors or a dystopian stainless-steel science fair project." (The diagram of the federal facility at Terre Haute, Indiana, where Timothy McVeigh, the "Oklahoma City bomber" was executed, provides a similar example. See Gillespie 2003: 93.)

Higgin et al. (2011: 175) describe the socio-technical organization of farm slaughter, offering a rubric of sorts for its 'choreography' that maps easily onto the choreography of prisoner execution just described. They outline the characteristic features as including co-ordination of *materials, bodies, spaces, timings,* and *enunciations* – this latter feature including the ways in which ideas, words, and sayings intervene in and 'order' practices.

> It is a matter of the co-ordination of *materials* [such as stun guns and stun baths, knives and automated cutting blades, restraining pens, hoists, electric prods, shackle lines, eviscerating machines etc.]; *bodies* [both in relation to the training of slaughterers and their highly specialized tasks but also the varied physical presences and agencies of animal bodies]; *spaces* [in terms of the physical layout and spacing of the abbatoir]; *timings* [e.g. the continuous speed of the overhead shackle line or the frequency of animal deliveries or the gap between stunning and 'sticking']; and *enunciations* [not just ideologies of killing as separate from practices of killing but the way in which ideas, words and sayings can intervene in and 'order' practices].

One difference we might observe between the slaughterhouse and the prison execution chamber, though, is in the *timings* just described. Animals are killed and processed as quickly as possible – without ritual or ceremony. The rapid speed of the slaughterhouse assembly line, factored by the many ostensible 'costs' of mistakes or delays, ensures what Wadiwel (2015: 163) describes as the optimal realization of value and profitability of the animal commodity. But the slow, tick-tocking passage of time experienced during the prisoner's death watch and execution is a torture unto itself for the condemned prisoner, and perhaps for many staff assigned the execution detail as well. The deliberate, protracted pace of the execution re-signifies again the higher value placed on human versus nonhuman animal life. But here we might recognize that while the different 'speed' of assembly-line killing in the prison and slaughterhouse is noteworthy, such differences only illustrate more poignantly the extent of carceral power to reach its endgame in ways it deems desirable.

Executions were ostensibly to become swifter, painless, and more efficient with advancements in 'modern' killing technologies – historically shifting from hanging, to the firing squad, gas chamber, electric chair, to today's lethal injection (Christianson 2010). All of these methods speak to the perverse underlying ambition of the 'precision of the correct death', however much the putatively advanced science or technology intrinsic to them is but a façade (e.g. Basu 2015). As prisoners' first-hand accounts attest, the process is anything but efficient, precise, sanitized, or technologically advanced; it is instead deeply saturated with drama and emotion – of both prisoner and executioner; and typified by mistakes, delays, and costs (capital punishment costs the state exponentially more than life sentences; Equal Justice USA 2016; Lynch 2000: 8–13).

Patterson (2002: 131) and others have also made the convincing case that any attempts to make the killing more humane have been primarily intended so that it is "less stressful to the killers" (also see Smith 2002: 55). Moreover, the more 'subhuman' ('animalized') the victims are made to seem, the easier they are to kill (Chapter 3). As Deckha (2010: 37) argues, "the routinized violence against beings designated as subhuman serves as both a justification and blueprint for violence against humans." The emotional, psychological, as well as physical torture of the process is undeniable, including that of the corrections personnel involved in the process. Cabana (1996: 188–189) describes working as a guard during an execution by gas chamber: "washing [his] body down with a garden hose was perhaps the sickest part of the whole thing."

The design and production methods of the slaughterhouse were the precursor to assembly line production itself, including that aimed at killing humans. Giedion reissued *Mechanization Takes Command* (1948; 2013) usefully outlines the historical roots and social impacts of European and U.S. 'scientific management' and mechanization of work, and illustrates the deep connections between the Cincinnati and Chicago slaughterhouse technologies with Henry Ford's mass production of automobiles, 1918–1939. Patterson (2002: 53–110), too, argues that the industrialization of animal slaughter – its technologies, speed, efficiency, and assembly line techniques – inspired Henry Ford in automobile production and in turn paved the way for the slaughter of humans in the Holocaust: 'the road to Auschwitz begins at the slaughterhouse'; and in fact some of the soldiers who worked in the death camps first worked in slaughterhouses. As J.M. Coetzee's protagonist Elizabeth Costello declares in Coetzee's acclaimed novel *The Lives of Animals*, "Chicago showed us the way; it was from the Chicago stockyards that the Nazis learned how to process bodies" (1999: 72; also see Rifkin 1992; Kim 2011: 317). The industrialized killing at the Chicago stockyards, as well as Henry Ford's automobile assembly-line production, informed Hitler's genocidal plans, and in fact Ford was a major backer of Hitler (see Christianson 2010: 86–88 for details of these men's relationship).[7]

Yet it is also important to follow Christianson's (2010) argument that the U.S. correctional industry – i.e. the United States Department of Justice – invented the gas chamber 'long before Hitler'. Christianson (2010: 237–252) lists the 594 U.S prisoners executed by lethal gas from 1934 to 1999. Ample evidence demonstrates the deep connectivity between the slaughterhouse and the execution chamber; both their materials and technologies originated from the U.S. Military Industrial Complex of World War I, and later developed into a United States and Third Reich collaboration. U.S. scientists developed the scientific, legal, and ethical rationale for the lethal gas chamber, and U.S. firms partnered with German corporations that provided the gas. Ultimately the U.S. federal government patented two models from Eaton Metal Products of Denver and Salt Lake City, ca. 1939, which became the world's leading designer and maker of gas chambers for prison executions (Christianson 2010: 6–8, 102–120). Earl C. Liston's patent application (Fig. 2.5), #2,172,168, was actually a 'double-seater' gas chamber. The patent illustrates the manner by

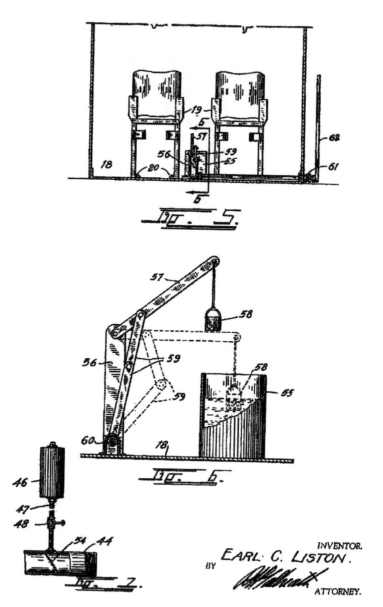

Figure 2.5 One of the drawings submitted with Earl C. Liston's gas chamber patent application, September 12, 1939.

which a mechanical device drops the cyanide pellets into a chamber, a process which "provides a neat, compact mechanism which will humanely execute the criminal or criminals with the least possible delay or confusion."

While arguments in support of constructing and patenting the lethal gas chamber focused on it as a 'more humane' method of killing compared with that of hanging or electrocution, as with every method of prison executions, lethal gas was eventually contested constitutionally in 1976 as cruel and inhumane. The last gas chamber execution was in Arizona in 1999. This is important to keep in mind since it is the ostensible illegitimacy of the *methods* of killing that have led to challenges of capital punishment, rather than the killing itself. All methods of capital punishment have been challenged juridically on the grounds of the Eighth Amendment to the Constitution's protection against cruel and unusual punishment (see Dayan 2007). This includes recent challenges to today's lethal injection drug cocktail (e.g. Supreme Court Upholds 2015). Perhaps the same could be said of industrial slaughter – challenges have rested on its inhumane ways of killing, not the sovereign or other right to kill itself.

Conclusion: the politics of sight redux

By way of a chapter conclusion I return to one of the most important aspects of the carceral sites discussed above: that is, their respective power to control and regulate 'visibility' of their operations. Since concealment, sight, and distancing are themselves crucial mechanisms of power and domination at carceral sites, to reverse these dynamics – to make their operations open and transparent – would seem to run counter to their carceral objectives (Berger 1980; Pollan 2002; Pachirat 2011; Gruen 2014b). As Pollan (2002) provocatively suggested, replacing the concrete walls of the concentrated animal feeding operation (CAFO) with glass walls would produce social and political transformation: "were the walls of our meat industry to become transparent, literally or even figuratively, we would not long continue to do it this way. Tail-docking and sow crates and beak-clipping would disappear overnight ... For who could stand the sight?" Merritt and Hurley (2014, above), too, argue that the invisibility of the operations of the prison and factory farm are required, otherwise, their "visceral nature would be publicly inflammatory."

The social, political, and legal maneuvering around the visibility of carceral settings raises many thorny issues. Finsen and Finsen (1994: 5–54) demonstrated that animal rights activists such as People for the Ethical Treatment of Animals (PETA) were able to prompt new protective regulations during the 1970s by making animal suffering, such as painful medical experimentation on monkeys, increasingly visible via media exposure. Visibility itself became a tool of advocacy, and disavowal impossible. But today, in our world of 'ag-gag' laws, instead we see those who fight against such concealment facing criminal charges if involved in the production, possession, and distribution of records of such hidden work (Kyle and Sewell 2015; see Chapter 4). As Rasmussen (2015: 64) reminds us though, the agricultural spaces regulated by ag-gag laws are not best understood as conflict between human and animal interests, but rather by the "demands of biocapital that establishes relationships of power between some bodies and others."

Timothy Pachirat, in his *Every Twelve Seconds* (2011: 233–256), develops a persuasive line of reasoning about why visibility alone is unlikely to produce social and political transformation, and I draw on it extensively here. One important aspect, as discussed previously, is that the routinization of the violence in assembly line killing makes it seem 'normal' and commonplace. As Pachirat (2011: 38–84) discovered, hiding most of the killing from vulnerable workers who are themselves exploited by the meatpacking industry ensured both their ignorance about industrial practices as well as a desensitization to them; even if he also found that the turnover rate of laborers at the Omaha slaughterhouse, for a number of reasons including hazardous working conditions and low pay, was 100 percent annually. But this volatility also poignantly raises the question of whether breaching these zones of confinement to reveal the tortures that occur inside would lead to their eradication. The answer is not at all that straightforward.

Many places of violence against animals – such as bullfights, rodeos, and zoos – serve as everyday human entertainment venues; zoos in Scandinavian countries today euthanize and grotesquely dissect 'superfluous' animals, including large mammals such as tigers, lions, and giraffes, in full view of an appreciative public (see Chapter 5). Being made visible – sight, exposure, or looking itself – will not ensure or guarantee any particular affective response on the part of the viewer and become a tool of animal advocacy. Kim (2011; 2015) illustrates this with examples across various cultural groups, whose practices range from 'the prosaic to the sadistic'. For example, in San Francisco's Chinese live meat market, seeing a live animal butchered is what consumers *want* to see; the live killing means taking home better, fresher, more nutritious, 'pure' food.

'Slaughterhouse tourism' popular in the 19th century provides another appropriate example. As Pacyga (2015) argues, such tourism was organized to convince consumers that the industry was producing good, high quality meat. Contra Pollan, the 'glass walls' of the Union Stockyards slaughterhouse did not produce disgust and horror at this spectacle of industrialized death. Fifty thousand workers, 'disassembling with breathtaking efficiency' 600 cattle, hogs, and sheep every hour became a must-see tourist attraction after it opened in 1865, drawing more than half a million visitors each year. While the tourist gaze was framed to neutralize and *naturalize* the horrors occurring inside the slaughterhouse (Pacyga 2015), today animal rights groups such as PETA routinely expose the horrors of animal abuse in order to *de-naturalize* them. Both might be considered propaganda, but we must ask, propaganda for what end?

Questions about the politics of sight, concealment, proximity, and so forth with respect to human prisoner executions presents a similar set of trending and complex dynamics. Gottschalk (2006) described the move from public to private executions, alerting us to the state's benefit in keeping hidden the horrors undertaken behind closed doors of the prison. But what might transpire were they to become transparent (again)? The vitriol of those who 'want to see him burn' is without question part of the discourse around capital punishment. Undoubtedly many sadistic individuals would want to *win* the hypothetical lottery that Pachirat proposes to require everyday citizens to perform state executions (2011: 241–243). Education about what was happening behind the closed door of the prison seemed to me at one time, the key. Surely if the public was made aware of what was happening in their name, I reasoned, their response would be repugnance (Morin 2013). But such is not the case; any particular affective response to these horrors is unpredictable. Bringing education to the public along with images of suffering to produce particular affective responses, such as is PETA's strategy, has to some extent been a successful strategy but is also not without controversy (Kim 2011). And many are simply indifferent to the pain and suffering of others, particularly of animals they do not understand, and of 'criminals' – who, after all, are just 'getting what they deserve' (Socha 2013; see my discussion in the next chapter about relationships across prisoner criminalization, racialization, and animalization).

All of which suggests, as Pachirat (2011: 249) argues in citing the work of sociologist Norbert Elias (2000) and others, that part of the civilizing project itself is the expansion of the 'frontiers' of repugnance (Sontag 2003; Deckha 2013b). Foucault (1977) exposed the deep contradiction in considering hallmarks of modern 'progress' the development of the ostensibly efficient, antiseptic, and technologically advanced carceral methods described throughout this chapter. Indeed, it is the concealment and displacement of violence, rather than its elimination, that is itself the hallmark of 'civilization' (Elias as quoted in Pachirat 2011: 249). Among other things, this cautions us against attempting to see any necessary correlation between 'seeing' and developing particular ethical or affective sensibilities; and, as Pachirat notes, profits can be made from offering the 'pleasure' of watching and witnessing horrific acts.

Rasmussen (2015: 56), too, troubles the assumption that 'seeing' something violent or troubling transforms peoples' ethical sensibilities or leads to particular forms of political action. In Rasmussen's examples of animal pornography, violence against animals is aestheticized and eroticized, such that the harm actually mobilizes desire for sexual gratification within the viewer rather than producing disgust. But as Rasmussen (2015: 66) also usefully points out, there is always a broader set of power relations and representational politics at work to produce particular affective responses. To understand the complex politics of visibility, or of any sensory reaction, we must consider the ways in which affect is not just an individual's 'private possession', but rather a "socially operated inscription of who is sufficiently like us in the ways that guide deliberations." This, she suggests, is a *political* rather than an ethical response to violence. How these reactions might be effectively addressed by critical human and critical animal geographers and others is a subject to which I return in the Afterword.

Notes

1 There is some indication that this number has dropped to 2.2 million today, partially owing to the release of many non-violent drug offenders from federal prisons. Still, if we include all those trapped in the PIC via probation or parole, the number overall is closer to 6.7 million (Williams 2016).

2 The Nebraska story is an interesting one. Just the name of the group that opposes the death penalty would seem to illustrate the tenuous political landscape they maneuver within: they formerly called themselves 'Nebraskans Against the Death Penalty' but today (using the same acronym) are 'Nebraskans for Alternatives to the Death Penalty.' Bosman (2015a; 2015b) reports on the controversy that erupted after the legislature repealed the death penalty in Nebraska in 2015. The wealthy governor, Pete Ricketts, drawing from his and his father's own bank accounts, began a petition campaign, Nebraskans for the Death Penalty, to reinstate it, and was able to secure enough signatures to place the issue as a referendum on the ballot in November 2016. The referendum passed (Nebraska Keeps Death Penalty 2016).

3 There were just 20 executions in 2016, compared with the peak year of 98 executions in 1999. Moreover, only 30 people were sentenced to death in 2016, compared with 315 in 1996 (Toobin 2016; Death Penalty Information Center 2017).

4 A similar and infamous case is that of Ricky Ray Rector, executed in Arkansas in 1981. Mentally impaired, he requested that staff save the dessert from his last meal 'for later' (Gottschalk 2006: 197).

5 More recently, as the Black Lives Matter movement has amply shown and resisted, some human lives are deemed to be worth less than others. The acquittal of Minnesota police officer Jeronimo Yanez in the shooting death of the unarmed Black man Philando Castile has dominated the news media recently (Smith 2017; Cobb 2017b). Castile's girlfriend had live-streamed and narrated the scene while seated next to Castile in their car's front seat, casting considerable doubt on the line of defense offered by Yanez's attorneys. The officer had demanded Castile's credentials, but proceeded to shoot Castile as he reached for them. Castile's death follows on the similarly 'innocent' police killing of hundreds of other unarmed Black boys and men over the last decade, including Michael Brown (in Ferguson, Missouri in 2014), Tamir Rice (in Cleveland in 2014), Eric Garner (in New York in 2014), and Freddie Gray (in Baltimore in 2015). These deaths across the country have sparked a national debate about police conduct toward Black people and within Black neighborhoods.

6 Although if one is looking, most transport trailers for livestock are unmistakable as the animals are visible through their open-air slats. Night-time transport is usually performed because the animals (especially birds) are calmer, there is less traffic, and the temperatures are cooler at night in the summer. Which is not to detract from one of my own memorable sights one December night a couple of years ago, driving along the highway in nearly zero temperatures, being passed by an open-air truck that was stuffed with hundreds of caged chickens, what appeared to be dozens to a cage, with their feathers flying about and a general sense of pandemonium more descriptive of the scene.

7 Meanwhile any number of related technologies emerging from the respective European and U.S. industrial revolutions aided in this association, perhaps none more surreptitiously but effectively as the invention of barbed wire. Netz (2004) offers a detailed discussion of what must surely be one of the most significant symbolic but also actual instruments of violence and spatial control within the prison as well as agricultural industries (1874 to 1954). Netz argues for the critical importance of examinations of such technologies across species lines – in the case of barbed wire, from the control of cattle during colonization of the American West to control of people in the Nazi concentration camps and Russian Gulag. "When we set out to

offer a history that mentions animals, we should understand that the history of animals is not merely an appendix, a note we should add because it is missing in our present traditional, human-focused history. Rather, the history of animals is part and parcel of history – that reality where all is inextricably tied together: humans, animals, and their shared material world" (Netz 2004: 11).

3 The prison as/and laboratory

Sites of trans-species bio-testing

Introduction

Putting aside the more familiar historical cases of medical and other testing on concentration camp prisoners – victims of the Nazi Holocaust or Stalin's Gulag – for most of the 20th century such testing on prisoners was common in 'regular' U.S. prisons. From the 1940s to the 1970s, alongside the tremendous growth in the health industry itself, U.S. doctors regularly injected prisoners with malaria, typhoid fever, herpes, cancer cells, tuberculosis, ringworm, hepatitis, syphilis, and cholera to test their effects (Washington 2006; Talvi 2002; Reiter 2009; Hornblum 1998). Scientists and physicians pulled out prisoners' fingernails and inflicted flash burns to approximate the results of atomic bomb attacks, and performed various 'mind-control' experiments. By 1972, the pharmaceutical industry was conducting more than 90 percent of its experimental testing on prisoners, and half of U.S. prisons had such research programs. By 1974, 70 percent of drugs approved by the FDA were tested on prisoners (Dober 2008).

While experimentation on prisoners occurred over most of the 20th century and continues into the 21st, it is worth recognizing the parallel structure of exploitation and killing of millions of nonhuman animals in laboratory research and experimentation in the U.S. each year. These animals, from mice to rabbits to primates, are bred for medical, pharmaceutical, cosmetic, and other product research and spend their entire lives in research spaces.

In this chapter I turn my attention to carceral relationships across the Prison Industrial Complex (PIC) and the Medical Industrial Complex (MIC), particularly as they intersect within 'Big Pharma', to examine their respective biomedical and other research spaces and their use of prisoners and animals as testing material. Practices within the MIC also align with those of the agricultural industrial complex in creating exploitable and disposable 'commodities' out of society's most vulnerable populations. Throughout this chapter I emphasize the spatial, structural, and operational forms of violence embedded within the MIC–PIC relationship, while staying attuned to the embodied experience of those who are subjected to experimentation – the experience of living out life in a lab cage. Although the oppressive conditions of living out life in a lab cage and being subjected to experimentation is not the same for the nonhuman animals and the human prisoners I discuss, as I noted in the previous chapter, they do not have to *be* the same in order to associate and juxtapose the industrial operations and patterns of violence that take place within them. I define these spaces as mutually 'carceral' (after Moran et al. 2017) based on the detrimental nature of each's confinement, the forms and types of research practices, and the violence enacted upon respective human and nonhuman bodies. Thinking through these institutions and spaces as mutually carceral allows us to consider not only their similarities but also the respective 'contributions' each ostensibly makes to scientific inquiry and the larger, profit-driven health industry.

I begin the chapter by overviewing the vast scale of the multi-billion-dollar medical industrial complex that exploits nonhuman and human prisoner populations in spaces of research and clinical trials, one among many of the proliferating sites, activities, practices, and professions that are part of the much larger 'health industry' itself. The tradeoffs, ethical debates, and secrecy behind these research practices must be weighed against arguments that such test subjects are 'required' for medical advancements and related technological progress. The very medicalization of such practices in the sterile, concrete lab cage – such as the enactment of medical protocols by researchers in white lab coats – rings acutely similar to the medicalization of prison executions, with the presence of physicians in the execution chamber ostensibly ensuring death with precision and without undue 'harm' (Knapp 2017). The effect of both is in making the violence taking place within appear somehow more appropriate, sanctioned, and legitimate.

Among the topics I raise in this chapter is that of relationships between carcerality and 'purpose breeding' that extends across both animal and racialized prisoner populations. Through a number of compelling studies such as Rebecca Skloot's *The Immortal Life of Henrietta Lacks* (2010), Harriet Washington's *Medical Apartheid* (2006) and *Deadly Monopolies* (2012), and James Jones' *Bad Blood* (1981), the historical record documenting abusive medical experimentation on Black people has reached mainstream audiences. The notorious Tuskegee experiments exploited the most vulnerable population of poor, illiterate African Americans during the post-war 'Gilded Era of Research', 1932–1972, wherein hundreds of men were injected with syphilis, unbeknownst to them, to study the progression of the disease. The mid-century turn to experimentation on mostly African-Americans captive in U.S. prisons continued this 'medical apartheid' (Washington 2006). I juxtapose these particular vulnerabilities with that of purpose breeding of nonhuman animals for food and biomedical research. Many might argue that incarceration and punishment of prisoners is fundamentally different than captivity and abuse within the slaughterhouse or research lab because prisoners are incarcerated as a result of some wrongdoing, whereas animals are not. And yet, as

I argue below, the entire apparatus of the prison industrial complex relies on an array of social, judicial, political, and economic policies to ensure criminalization of Black bodies (Orson 2012; Cacho 2014; Kim 2017; Alexander 2012; Gilmore 2007; Glick 2013) that are – effectively then – 'purpose bred' for prison, and thus their vulnerabilities within the medical research apparatus is more closely linked to that of nonhuman animals than might appear at first glance. This vulnerability also requires, as I discuss below, assessing the extent to which captive and incarcerated individuals can ever be free to consent to such testing.

Much of my discussion below also focuses on how human–animal distinctions themselves are made (and potentially unmade) through carceral processes, a key proposition and concept underlying the book as a whole but one that has particular salience with respect to the use of nonhuman and human prisoner bodies as research material. 'Animalization' and 'humanization' are processes that place certain nonhumans and certain humans into hierarchies of worthiness and value. The social construction of 'the animal' or the subhuman other – that is, the derogatory social meanings attached to various groups – has been invoked to perpetuate hierarchical human–human and as well human–nonhuman relationships (Kim 2011; 2015: 24–60; Deckha 2010; 2013b; Cacho 2014; Jackson 2013; Glick 2013). In the process both certain human and nonhuman bodies become valued and protected, while others become vilified, reduced to the status of beastly animal; and thus providing the carceral logic that governs which bodies can be exploitable, disposable, and/or killable in the research lab. Understanding the process of animalization of humans and nonhumans, is, to me, one of the most compelling analyses taking place within critical animal and critical race studies today. As I explore below, the carceral logics underlying domination of numerous vulnerable populations intersect and are deeply intertwined (Deckha 2013b: 515).

Finally, the chapter's conclusion touches on the legal frameworks that allow humans and animals to be used as experimental material for profit. Biomedical research on prisoners and animals represents an array of overlapping regulations that exclude most prisoners and animals from important legal protections; they are haphazardly upheld; and they apply, in any case, only to scientific research funded by the federal government.

The Medical Industrial Complex: Big Pharma and trans-species bio-testing

The Medical Industrial Complex shares many of the spatial, structural, operational, and effective outcomes of the Prison- and Agricultural Industrial complexes, as well as sharing many of the same types of relationships across government and industry. The testing of pharmaceutical drugs and other products in the research laboratory is part of a much larger 'health industry' and apparatus of the Medical Industrial Complex (MIC 2017) that includes a multi-billion-dollar conglomerate of doctors and medical schools, hospitals, nursing homes, insurance companies, drug manufacturers, hospital supply and equipment companies, real estate and construction businesses, health systems consulting and accounting firms, attorneys who file medical malpractice suits, manufacturers, retail pharmacies, and banks. As Ehrenreich and Ehrenreich (1971) argued decades ago, the concept of the Medical Industrial Complex (or "American Health Empire") conveys the idea that the primary function of the health care system in the United States is business – that is, to make profits – with two other secondary functions, research and education (Dober 2008; MIC 2017). Among the panoply of institutions and actors that make up the MIC, the key institution producing an ever-increasing portion of health care profits in recent years is that of "Big Pharma," pharmaceutical companies that undertake research on drugs and other products via a vast network of scientists and laboratories at research universities, federal government agencies such as National Institute of Health (NIH), and other private and public companies and facilities. The billion-dollar scale of Big Pharma's profits requires serious attention, relying as it does on the incarcerated bodies and carceral spaces and practices that make such profits possible.

The close of the 20th century witnessed skyrocketing profits in prescription drug sales in the U.S., increasing 500 percent from 1990 to 2005, comprising at that time a $200.7 billion-dollar industry. By 2016 sales had more than doubled again, increasing to $497.5 billion. Global pharmaceutical sales topped $1 trillion in 2014. This made the worldwide drug industry the second most profitable industry overall, with a 19.6 percent profit margin in 2014, which compared to 6.9 percent profit margin for all Fortune 500 companies. (According to Forbes, by 2016 it had become the most profitable industry overall.) The world's 10 largest drug companies generated $429.4 billion of that revenue (IMS Health 2017; Health Costs and Budgets 2017; U.S. Department of Health and Human Services 2017). In 2014, those top 10 pharmaceutical companies (by sales) were: Johnson & Johnson ($74.3 billion); Novartis ($57.9 billion); Roche ($49.8 billion); Pfizer ($49.6 billion); Sanofi ($43 billion); Merck ($42.2 billion); GlaxoSmithKline ($37.9 billion); AstraZeneca ($26 billion); Bayer ($25.4 billion); Gilead ($24.4 billion); with four of these originating in the U.S. (Johnson & Johnson, Pfizer, Merck, and Gilead).

Among other things, the staggering profits of these companies allows for the purchase of 1,100-plus paid lobbyists who exercise powerful influence on federal lawmakers in Washington D.C. From 1998 to 2014, Big Pharma spent nearly $2.9 billion on lobbying expenses – representing more lobbying effort than any other industry. The industry also spent (for example) more than $15 million in U.S. political campaign contributions from 2013 to 2014 (Big Pharma 2016).

Pharmaceutical companies spend billions of dollars annually on drug research, development, and advertising. Private charities and foundations account for only 5 percent of the estimated $100 billion spent on biomedical research in the U.S. each year; pharmaceutical and medical device companies contribute approximately 60 percent. As Dober (2008) argues, these doctors, scientists, research organizations, medical journals, teaching hospitals, and university medical schools all exhibit disturbing conflicts of interest between their publicly stated missions and their financial and ideological subjection to Big Pharma. Because drug companies sponsor clinical trials that researchers are paid to administer, too often the academics and scientists supply subjects and collect data according to the instructions and needs of their corporate sponsors, rather than answering questions that might otherwise emerge from independent basic scientific research. (This is just one example of the increasing 'corporatization' of the university about which scholars and scientists, including geographers, have raised concerns; see e.g. Castree and Sparke 2000.) Sponsors typically keep the data, analyze it, write the papers, and decide whether, when, and where to submit them for publication.

It can take a decade or more between a drug's initial development and its approval. As Dober (2008: 7) describes, it is not unusual for drug companies to own subsidiaries in the cosmetics and 'lifestyle' drug industry, which earn the biggest profit margins by addressing conditions such as erectile dysfunction, baldness, skin wrinkles, and obesity. On the other end of the spectrum, because competition is so fierce, the federal government provides incentives for companies to seek drugs and cures for less common diseases. In such cases the company will receive major tax credits for producing them and then be offered market exclusivity protection for seven years, provided by the Food and Drug Administration (FDA). The overall effect is even larger profits, funded by the American taxpayer. In other words, government incentives mean that U.S. law allows drug companies to set the prices for drugs and then protects them from free-market competition. The true cost of developing a drug is 'shrouded in mystery' with many unverifiable figures reported by Big Pharma; some watchdog groups argue that drug companies' true cost of producing new drugs is approximately half of what they report to the government (Light 2017; Dober 2008: 8). To take just one example, since no competitors were allowed to develop a generic version of the AZT treatment for the HIV virus, the federal government allowed the drug company Burroughs to charge $8.00 a dose for a drug that cost only .63 cents to produce. To add to this scenario, privately funded research is not subject to government regulation, and is only expected to grow in years ahead.

Current trends in the drug industry's enormous growth and profit fuels the demand for trial subjects. Dober (2008: 7–8) refers to test subjects as valuable "commodities … [that need to be] affordable, available, and useful. Currently, like any commodity in short supply, research subjects are not affordable or readily available." While using prisoners in biomedical and other research has come under greater governmental scrutiny compared to the mid-20th century, as I discuss below, it continues today with little oversight (Reiter 2009). And of course using animals to test medicines, pharmaceuticals, and cosmetics has reached unprecedented levels in today's commodity markets.

Commodifying vulnerable populations

Embodied geographies in the lab cage: prisoner lives

Biomedical and other research involves three phases of trial before their results can be sold for public consumption. Pre-clinical or Phase I clinical trials represent the first time that new drugs or treatments are tested, typically on small groups of animals or sometimes people, to determine their efficacy and safety and before testing on larger human study groups in Phases II and III. Although officially banned by the Food and Drug Administration in 1980, evidence confirms the continuing use of human prisoners as research subjects for Phase I clinical trials of previously untested drugs and products.

The late 1940s and 1950s saw huge growth in the U.S. pharmaceutical and health care industries, accompanied by a boon in prisoner experiments funded by both the government and corporations. In fact, the United States was the only nation in the world to officially sanction the use of prisoners in experimental clinical trials (see Figs. 3.1 and 3.2). These practices were not dissimilar to those of the Third Reich occurring at the same time – practices condemned in the high-profile Nuremberg Trials, when U.S. prosecutors were seeking death sentences for Nazi doctors accused of unethical medical research (Washington 2006: 232; Dober 2008: 6). In fact, as Reiter (2009: 508) points out, Nazi doctors defended their atrocities by citing similar practices taking place in U.S. prisons.

One of the most notorious cases involving prisoners in medical research ended with a lawsuit in 2000, when 300 former prisoners filed suit against University of Pennsylvania dermatologist Albert Kligman, Dow Chemical, Johnson & Johnson, and the city of Philadelphia, for experiments conducted at Philadelphia's Holmesburg Prison between 1951 and 1974. Hornblum (1998) documents the mostly low-income African-American prisoners who were paid $2.00 or $3.00 dollars per day to test dangerous tranquilizers, dioxin (the main ingredient in Agent Orange), powerful skin crèmes, new cosmetics, and other chemical agents. Washington (2006: 250) discovered that Kligman went well beyond dermatological experiments to inoculate men with syphilis, gonorrhea, malaria, and amoebic dysentery, as well as testing psychotropic drugs for the Army and the Central Intelligence Agency (also see Reiter 2009; Stobbe 2011; and Thompsons's 2016 work on Attica prison experiments). Survivors suffered all number of problems including skin problems, breathing difficulties, cancers, unidentified infections, and psychological disorders.

Hornblum (2007) examined in detail the experience of Edward Anthony, one of the thousands of prisoners at Holmesburg to suffer from Kligman's laboratory tests. Hornblum's chapter, "My Back Is On Fire," recounts Anthony's first nights and weeks of being subjected to a "bubble bath" test (2007: 1–7). The doctors had removed layers of skin off of Anthony's back and then coated it with searing chemicals. He became dizzy and nauseous, and recalls, when back in his cell,

Figure 3.1 In 1966, Solomon McBride examined an incarcerated subject wearing skin patches coated with various experimental pharmaceuticals in H block of the Holmesburg Prison complex. Reprinted with kind permission of the Urban Archives of Temple University, Philadelphia, Pennsylvania.

> Man, my fuckin back is on fire. It's killing me … my back was screaming. I really felt like my back was on fire.

Other prisoners observed large blisters the size of nickel coins on Anthony's back that had filled with pus. He recalled,

> All that day and in to the night it starts getting worse. It was unbearable. My whole fuckin body is reacting. It was like something was crawling under my skin. Under my arms and between my legs it's getting real hot. I'm moaning ….

The next morning Anthony found red and white bumps and blisters all over his face, arms, legs, head, and chest. At that point he was taken off the clinical test, but doctors refused to see him or give him anything for the pain. His exclusion from the dining hall meant his meals were delivered to his cell, with other prisoners "looking at him as if he had something communicable."

> Back in my cell I'm still in pain. It feels like something is crawling under my skin. The itching was terrible. My armpits and groin are startin' to heat up and be on fire like my back. I couldn't keep my arms down by my side and I'm having trouble walking. Any sort of movement hurts.

Figure 3.2 In 1945, Army doctors exposed patients to malaria-carrying mosquitos in the malaria ward at Stateville Penitentiary in Crest Hill, Illinois. Photograph by Myron Davis, *Life Magazine* @Time, Inc.

Anthony suffered weeks of intense itching and agonizing pain. As Hornblum asserts, Anthony "was not alone – there were thousands of desperate, incarcerated men and women just like him in postwar America. Imprisoned Americans … shockingly discovered that they had not only been sentenced to prison, but sentenced to science as well" (2007: 7).

Washington also argues that Holmesburg was "no anomaly" either (2006: 252–253). Physicians from the Sloan-Kettering Institute injected prisoners in Ohio's state prisons with live human cancer cells in 1952. In Alabama prisons from 1967–1969, testing of plasma transfusions resulted in widespread hepatitis. Other prisoners across the country were subjected to flash burns – exposure to heat radiation – including to the cornea of the eyes (Washington 2006: 253). At Tulane University in 1950, electrodes were implanted in prisoners' brains to repeatedly stimulate their pleasure centers. And at Washington State prisons from 1963–1973, prisoners' testicles were radiated and dissected (see Stobbe 2011 and Clark 2014 for many additional examples).

As Washington (2006: 256–257) alerts us, until the 1970s news coverage of prison research was almost universally laudatory, celebrating prisoners who volunteered for dangerous experiments that would benefit society while at the same time making 'real' money while incarcerated. A number of lawsuits over the decades have been filed in order to *allow* medical and drug testing on prisoners, though, and a number of prisoners and class action lawsuits have been filed to redress harm done by them, or, by contrast, allow them to freely participate if they wish. In 1979 prisoners from the State Prison of Southern Michigan, for example, filed suit to prevent the FDA from excluding them from research studies. Talvi (2002) and Washington (2006: 267–268) discuss Brown University's 1999 attempt to allow testing on HIV-positive prisoners, prisoners who were not, however, given the 'known' cures of the time, but rather experimental drugs that carried dangerous medical risks. The Brown University argument rested on the vague 'right' of prisoners to such experimentation.[1]

During widely covered congressional hearings in 1973, the pharmaceutical companies acknowledged that they were using prisoners for testing because they were "cheaper than chimpanzees," due to the increasing shortage of primates available for biomedical studies (in particular, rhesus macaques could no longer be imported from India; Dober 2008). Works by journalist Jessica Mitford (1973) influentially exposed to a wide public audience the massive exploitation of U.S. prisoners in pharmaceutical and other biomedical research (also see Landa 2009; Shubin 1981). Such experimentation was ostensibly put to a halt with a series of subsequent congressional regulations and the 1981 "Common Rule" that placed limits on such research, securing federal jurisdiction over medical research involving prisoners, and creating the Institutional Review Board (IRB) process. The IRB is intended to analyze risks and ensure protections on any research involving vulnerable populations, including prisoners. Despite these protections, prison research picked up again in the 1980s, and in 2006, the Institute of Medicine (IOM), which advises the federal government on biomedical issues, appointed a committee to study the issue and make recommendations – and they recommended that prisons be re-opened to more extensive medical research (Hornblum 2007: xiv; Washington 2006: 268; Reiter 2009; see Conclusion below).

Today, the American Correctional Association generally prohibits use of prisoners for medical, pharmaceutical, or cosmetic experiments unless the research fits within the permissible categories of "benefitting" prisoners somehow – the determination of which is, not surprisingly, highly controversial. There are serious ethical problems with suggesting that captive prisoners could ever be 'free' to consent to or volunteer for such testing (and of course animals are categorically unable to consent; see below). Yet, Maron (2014) and others argue for the rights of prisoners to offer themselves as test subjects. Maron substantiates her position by arguing that clinical trials need "better representation" of Blacks and Hispanics (who disproportionately comprise the prison population) because they disproportionately suffer from certain health problems such as Type 2 diabetes yet are not proportionately represented in clinical trials. Cohen (1979) concludes that it would be paternalistic to not allow prisoners to make their own decisions whether to participate in such research.

Yet prisons are also populated with poor, under- or un-educated individuals, individuals with mental illnesses, or those with alcoholism and drug abuse problems – approximately 80 percent of prisoners fall into one of these categories. The issue of compensation, too, is difficult to resolve in the prison environment where "even small payments or amenities are significant inducements" (Dober 2008: 12) – e.g. to people with tiny bursary accounts attempting to purchase exorbitantly over-priced commissary or vending machine products. Moreover, volunteering for medical research was and often continues to be the only means of accessing competent medical care. Because diseases of the poor disproportionately afflict prisoners, they make 'good' test subjects; approximately 60 percent of imprisoned Black men have hepatitis C (U.S. Department of Health and Human Services 2017). The prisoners at Holmesburg had 'volunteered' and signed consent forms, yet most likely did so either because they did not understand what the vague, confusing language of consent forms actually meant, or did so in order to be granted special privileges, to be viewed positively by prison personnel, or to receive better medical care (Washington 2006: 258–264). While HIV-positive prisoners might be pressured into enrolling in drug testing, for example, the serious lack of medical care and treatment in prison might easily incentivize prisoners to participate even if testing could expose them to serious medical risks. It bears repeating: because health care provision is so poor to nonexistent in prisons, many prisoners may see participating as their best avenue to any care at all.

Talvi (2002), Stobbe (2011), Clark (2014), Reiter (2009) and others warn that medical testing on prisoners is on the rise, what Washington (2006: 266) calls a "quiet renaissance" in prison research (meanwhile Big Pharma is also moving much of its testing 'offshore' to other vulnerable and poor populations). Evidence suggests that since the 1980s, prison-based research is again being conducted in numerous states including Arkansas, Connecticut, Florida, Maryland, Rhode

Island, South Carolina, and Texas, in clear violation of federal laws (most are conducted under the auspices of the Department of Health and Human Services). Reiter (2009: 502) reports that from 2006 to 2008 the drug company Hythian conducted experimental, unapproved drug addiction treatment programs in the states of Washington, Indiana, Texas, Louisiana, and Georgia. Other experiments induce labor in pregnant prisoners, test methods of obtaining biopsies, and test experimental drugs and vaccines for HIV, hepatitis C, and cancer (Talvi 2002; Washington 2006: 268–269). One excessively invasive experiment for the treatment of lung cancer involves heating a prisoner's blood and putting it back into his body and artificially creating hours of hyperthermia – and meanwhile putting him at risk for seizures, congestive heart failure, burns, heart attacks, limb loss, and death. Dober (2016) offers numerous examples of current drug testing on prisoners, including 'free' use of prisoners for brain imaging research in Wisconsin and New Mexico, as well as the Houston police department using prisoners for the training of officers to draw blood and thus expedite drunk-driving arrests.

Such examples prompt us to ask, who is ultimately set to benefit from such researches? Most likely not prisoners themselves; mainly the drug companies benefit, as well as perhaps those consumers able to afford access to new pharmaceuticals on the market. One must also pose questions about who can best represent prisoners on IRB boards. Today, correctional personnel or drug company employees are those in position to frame the risks to the IRB panels, but their assessments may quite possibly have other primary motives than protection of prisoners. And as Clark (2014) argues, as "human guinea pigs," the vulnerability of human prisoners is something they share with their nonhuman counterparts. Just with respect to the language, discourse, and vocabularies describing the corporeal 'care' of caged bodies in the laboratory, the same words – feeding, cleaning, grooming, labeling, testing, exercising, and tracking – are used for both humans and nonhumans (Leder with Greco 2014). It is to the lived experiences of those nonhumans in the lab cage I turn next. (And in Chapter 5 I pick up the discussion of commodification of vulnerable populations in broader terms; that is, in the processes by which prisoner and animal bodies become saleable and exchangeable as property and commodities in a wide range of carceral spaces.)

Embodied geographies in the lab cage: animal lives

Much debate surrounds the use of animals in biomedical and other laboratory experimentation, debates that hinge on the putative beneficial trade-offs in knowledge gained for human wellbeing.[2] As one of the headlines for the Foundation for Biomedical Research (2017) declares, "Lab animals have made important contributions to nearly every Nobel Prize in Medicine." The 'biopolitical' state's ostensible mandate to protect society and foster life (Taylor 2013: 542) thus seems to translate seamlessly into the right to sacrifice lab animals for human benefit. But even if this position were somehow defensible, it is the case that most animal testing is superfluous, intended mainly for increasing corporate profits. Not mentioned in the discourse of saving or improving human lives are the billions of dollars the drug and cosmetics ('beauty') industries earn in selling the vaccines, chemical substances, and other products derived from animal bodies. After more than two decades following a 'no testing on animals' policy, for example, cosmetics companies Avon, Estée Lauder, and Mary Kay have just resumed the practice. And to take another representative example, at a federally funded (and remote) research center in Nebraska, Moss (2015) found callous tests being performed on farm animals that were intended simply to breed meatier offspring and thus create higher profit margins for the beef, pork, and lamb industries. In the process, however, thousands of animals were left to die of starvation, deformities, and other such ailments and weaknesses.

As these brief examples illustrate, the utilizing of animals for research encompasses a broad range of activities taking place in a wide range of sites, most of them hidden from public view. These range from the testing of drugs, vaccines, medicines, cosmetics, procedures, devices, and other consumer products at laboratories located in public and private institutions and corporations; to observing animal behavior in situ and ex situ; to using animals in education and training (e.g. dissection in medical school classes); among many others.

Approximately 17 million nonhuman animals are used for laboratory research and experimentation in the U.S. each year (Gruen 2011: 107; she estimates the global number to be 115.3 million a year, 100,000–200,000 of whom are primates).[3] Such animals are bred for research and spend their entire lives in research spaces. Animals used in laboratory tests range from small vertebrates such as mice, moles, and rats; to amphibians; to rabbits and ferrets; to cats and dogs; to large primates and monkeys. Examples of such testing includes rats, dogs, and primates used for testing of toxicity of drugs; rats and mice used to study cancer-causing effects; and rats, mice, and rabbits used to test the absorption, distribution, metabolism, excretion, and interactions of active drug ingredients (see Figs. 3.3 and 3.4). There are 3,337 strains of mice alone purchasable for research. Urbanik (2012: 79, 86–89) offers a valuable discussion of genetically engineered mice. Mice have been bred in the U.S. for research purposes since 1929; they make 'good' research subjects because they reproduce rapidly, can be cared for inexpensively, handled easily, and their genetics can be easily manipulated. Davies (2013), though, alerts us to some of the biggest questions surrounding these 'model' organisms, including how inbreeding and mutancy has affected their species identity and individual experiences.

Animals used in laboratory experiments are typically kept in sterile concrete or wire cages; they often suffer either from the isolation and boredom of captivity or the stress of crowding; and when not undergoing stressful, painful, and sometimes tortuous experiments, suffer the stress of anticipating them (Finsen and Finsen 1994; Gruen 2014a; Urbanik 2012). Acampora (2006: 91–115) refers to the laboratory cage as a site of "carcerality reified." The world of bars, walls, or fencing phenomenologically assimilates the "carceral into the carnal" and prevents a full range of corporeal expression on the individual, social, and

Figure 3.3 Standard lab cages with added paper as a rodent "environmental enrichment activity." *The Critter Cave Column*, 2010.

ontological levels (2006: 98–101). Particularly at the *ontological* level, rats and mice – by the numbers the most frequently used lab animals – are reduced to 'generic animals, eventually becoming only lab-functions (to the experimenter) or just inert commodities (to their suppliers)'. In this process, the organism "has become transformed from what most of us would commonly call an animal into something that stands in for data and scientific analysis … reduced to particular gene effects or physiological responses." For scientists to do their work, the essence and identity of animals must disappear (Acampora 2006: 102; Birke 2003). (I return to this distancing effect in the Conclusion.)

As with the slaughterhouse and execution chamber (Chapter 2), research laboratories are typically 'invisible' to a public that might otherwise question

Figure 3.4 Rabbits confined for cosmetics testing. Photograph courtesy of Brian Gunn, International Association Against Painful Experiments on Animals, n.d.

the practices taking place within them. The high degree of secrecy involved in their methods has very likely prevented more public and government involvement in their regulation than has been evident. Orzechowski's *Maximum Tolerated Dose* (2012) offers a particularly poignant and haunting filmic portrayal of the secrecy surrounding the spaces of animal experimentation. Like those of the slaughterhouse and execution chamber, laboratory protocols and methodologies also rely on the perverse notion of the 'precision of the correct death' with highly medicalized, regulated, and surveilled spaces and practices, carefully documented for publication and reproduction of testing results. Gruen (2011) further observes that these results are moreover written in the most obscure scientific language to hide the procedures and suffering that goes on – making the language itself sanitized by referring to animals by numbers or other labels.

DeMello (2014: 79) investigates another commonly used lab animal in the U.S for biomedical research and cosmetic product testing – rabbits. Rabbits have been bred specifically for laboratory use since the mid-20th century, and large suppliers provide millions of animals per year for such purposes. As DeMello describes,

> rabbits typically lead lives of isolation. Because most are not surgically sterilized, they are kept alone in small, steel cages to prevent fighting

and unwanted reproduction, and typically have nothing to play with and nothing to do. Rabbits, like other laboratory animals, are often observed engaging in stereotypical behaviors associated with emotional and psychological deprivation, such as bar licking, excessive grooming, or paw chewing, and sitting in a hunched position for hours at a time ... In addition, they often either undereat or overeat to counter their boredom; and many develop deformities in the spine and legs because they can't move freely in tiny spaces.

Cages for these rabbits are too small to permit normal behaviors such as sitting up on hind legs, hopping, digging, and hiding. DeMello adds (2014: 83) that those who are eventually moved to rescue facilities often have "enormous psychological problems stemming from their lack of social contact and intellectual stimulation" as well as physical problems stemming from intensive confinement.

Turning to larger species, the NIH conducts primate research at seven large sites throughout the U.S., including the Yerkes Center in Atlanta. Among other practices there, squirrel monkeys are routinely subjected to cocaine addiction and malaria. Because these animals are directly injected with infectious diseases, their housing – again, which they inhabit for their entire lives, which can be decades – has become ever more sterile and controlled, with sealed concrete walls and floors the norm. From the 1920s until very recently, Yerkes had been conducting research on captive chimpanzees. Though extensive requirements for nutrition, space, husbandry, and veterinary care for chimps were arguably part of Robert Yerkes' original plan for the facility, the chimps lived impoverished social and physical lives, removed from their mothers at birth, receiving chest tattoos with their institutional record number – institutional markings not unlike those of the incarcerated prisoner.

Ross (2014: 65–67) reports from the 1940s into the 1980s, Yerkes was the premier site for contaminating chimpanzees with human infectious diseases, including leprosy and the AIDS virus. Again, the nature of these diseases as infectious dictated particular sterile and controlled laboratory environments. But by 2002 public pressure led to changes in federal regulation over chimpanzee production and care such as those requiring that an 'ethical environment' be provided (see below), and establishing a sanctuary system. By 2011, chimpanzee experiments had been banned in every other country in the world except the U.S. (Gruen 2011: 129); today such research is coming to an end with only 0.056 percent of active NIH funded projects using chimpanzees in invasive biomedical research. Today chimpanzees live out their lives in a 'dizzying array of housing circumstances' depending on why and where they were bred in the first place; most though, remain captive within 'unnatural' physical settings subjected to long-term harmful effects (Ross 2014: 55, 67–68; Gruen 2016b).

Zellhoefer (2013: 244–247) draws our attention to the breeding of cats and dogs for vivisection, and the many experiments on products that do not require FDA approval. Animals and humans share only 1.16 percent of diseases, so as

Zellhoefer argues, what works on animals will not necessarily work on humans anyway, and moreover drug testing on animals has a 92–95 percent failure rate:

> Ironically and disturbingly, a common justification for the continued use of animals in research is that they are so much like us we can learn from them, yet at the same time, researchers claim that animal testing is justifiable because human animals and nonhuman animals are fundamentally different (2013: 246).

Compelling arguments against laboratory testing on animals come from those who represent the standpoint of animal suffering, as well as those like Zellhoefer who question research protocols and the efficacy of research results (PCRM 2017; Gruen 2014a). Vaccine testing alone kills 2.5 million animals every year. Abnormal toxicity testing, neurovirulence testing, potency testing, and 'challenge' tests all result in pain, suffering, and the deaths of animals injected with deadly viruses. To test the potency of a single batch of the rabies vaccine, for example, the live rabies virus is injected through the skulls and directly into the brains of mice (not the normal route of infection), some of which have been given the protective vaccine first. About half of the animals develop and/or die of rabies, a painful neurological disease involving tremors, loss of control over one's body, the inability to swallow, and severe weight loss. Yet the NIH itself admits that 92 out of 100 drugs that successfully passed such animal trials failed during the human clinical trial phase (NIH 2017). Today as the PCRM (2017) outlines, there are many cutting-edge alternatives and technologies to using animals in clinical research. These include increased use of human cadavers, bench-top simulators of various human organs, cell-based tests, and computer programs that can simulate medical devices, among others.

Finally, as I noted above, there are a number of instrumentalist arguments and ethical complexities surrounding prisoner informed consent to laboratory experimentation. But consent with regard to animals is quite a different matter. The idea of 'avoiding coercion' – as IRB protocols require for humans – makes little sense for animals. As Gruen (2011: 128) points out, animals simply cannot give informed consent, and moreover if they could, "we could certainly imagine them objecting to being held in a cage, in isolation, being subjected to invasive procedures, being denied opportunities for exercise and other capacities constitutive of their wellbeing." She notes that the laws governing the use of human embryos are more stringent than those for use of animals.

Carcerality and 'purpose breeding'

The nonhuman animals discussed above are extensively and routinely "purpose bred" for research; Urbanik's mice (2012) and DeMello's rabbits (2014) are brought into existence solely to be used as material in laboratory experimentation, and live out those existences in lab cages, unnaturally isolated from or crowded among other beings. In considering life under such carceral conditions, it is useful to pause and consider what Acampora (2006: 108) has referred to as "susceptibility to incarceration" of respective populations of humans or non-humans. That is, while we may observe resonances across the vulnerabilities, suffering, and systems of abuse within the carceral spaces of animals and prisoners, he submits that 'susceptibility' to being incarcerated can remain invisible. How do those respective human and nonhuman populations end up in these carceral settings in the first place? What are the respective processes and carceral logics that lead to their captivity and incarceration?

Many might argue that incarceration, punishment, and exploitation of prisoners is fundamentally different than captivity and abuse within the animal slaughterhouse or research lab because prisoners are incarcerated as a result of some wrong doing, whereas animals are not, they are wholly 'innocent' creatures deserving of our sympathy. Prisoners by comparison are 'getting what they deserve'. Many abolitionist-oriented animal rights scholars such as Steiner (2016) would argue that no caging or confinement of nonhuman animals can be justified, whereas at least some confinement (and accompanying carceral practices) of human beings can be.

And yet, as many scholars of carcerality have found (Gilmore 2007; Alexander 2012; Wacquant 2001; 2005; 2009; Peck 2003), the entire apparatus of the Prison Industrial Complex relies on a steady supply of bodies to perform socially-constructed 'crimes' in order to keep the neoliberal security state and market economy – especially at the state and local levels – functioning. Peck (2003: 226) contends that the reductions in the welfare state from the 1980s onward have been used to legitimate and harden new regulatory regimes and new forms of governmental rationality, suggesting that "the prison system can be understood as one of the epicentral institutions of these neoliberal times." Within the ostensible free market of state economic restructuring, cuts in welfare and other resources for the poor – ideologically as well as financially – were reallocated to the development of various policing and security apparatuses, including prisons, and thus creating a 'penal state' to stem the consequences of rising destitution (Jones 2010; Wacquant 2005; 2009). Those who are "structurally irrelevant" to capital accumulation are warehoused in prisons. This represents a punitive approach to social marginality, an "uneasy marriage between economic liberalization and authoritarian governance" (Peck 2003: 225; Hallsworth and Lea 2011: 142). (In Chapter 4, I discuss how those bodies can then become disposable commodities of public and private enterprises.)

These insights seem particularly germane when considering the structural racism that is a basic constituent part of the U.S. prison system. Alexander (2012), Orson (2012), Wolfers et al. (2015) and many others argue that mass incarceration in the U.S. itself is a racialized strategy for criminalizing and thus regulating (warehousing) the current 'surplus' Black urban poor population. Of course, how such 'surplus populations' are created or controlled varies with the circumstances and demands of the carceral or security state apparatus. As Washington (2006: 205) points out, for example, women under the plantation slave system had been forced to procreate, but under the 1980s dismantling of the welfare state, they were being forced into sterilization.

The PIC of today relies on a whole host of social, judicial, political, and economic policies to ensure criminalization of Black bodies that are – effectively – 'purpose bred' for prison. This has occurred most recently through the decades-long War on Drugs that targeted Black activities while ignoring or diminishing those of Whites. (Before that, the Counter Intelligence Program COINTELPRO, an FBI operation dating from 1956, secretly monitored and targeted, among others, the activities of Martin Luther King.) Beginning in the early 1980s and ostensibly in response to the crack cocaine 'epidemic', the Reagan administration escalated the drug war that had begun under Nixon. As Alexander (2012: 51–54, 112–114) relates, during this period the U.S. judicial system exaggerated the dangerousness of crack cocaine, casting possession of even a small amount of it as a far more serious crime and social ill than possession of pounds of powder cocaine – the drug of choice for White suburbanites. With the government's unprecedented expansion of law enforcement activities in poor urban Black neighborhoods, this meant a radical increase in the number of arrests for possession or distribution of crack cocaine. These were adjudicated as felony offenses and carried longer prison terms than those for powder cocaine (the penalty disparity was 100:1 by weight). (In addition to the racist logic of this disparity, this comparison also alerts us of the "libidinal economy" of Blackness that Wilderson 2010 discusses.) Meanwhile in 1984 the U.S. Congress also passed the Sentencing Reform Act, which abolished parole for federal prisoners and guaranteed that they serve at least 85 percent of their prison sentences, which predictably resulted in a number of problems such as prison overcrowding (and hence, justification to build more of them).

Some consider the mass incarceration of people of color a form of 'social death' and, ultimately, the genocide of a population – if not actually dead, they are the "walking dead" (Butler 2009; Guenther 2013; Cacho 2014). "Living, they nonetheless appear as if and are treated as if they were dead" (Gordon 2011: 10). The philosopher Claudia Card (2003) has defined genocide as the infliction of social death upon a particular population, a definition that has pertinence especially to Black communities today (also see Patterson 1951). One in three Black men will end up in prison at some point in their life – keeping alive the prison industry while destroying families, neighborhoods, communities, and entire ways of life. And again, while the social death of animals through a lifetime of captivity and the putative genocide of Black Americans through mass incarceration are not the same thing as annihilating a national, ethnic, racial, or religious group out of hatred (cf. Patterson 2002), they do not have to be the same in order for us to consider the effective outcomes and "indifference that helps perpetuate such conditions and losses" (Socha 2013: 223–229). To Socha this is the difference between hatred and indifference, and it is indifference that we need to address. Moreover, as Guenther (2013: xvii) argues, while Foucault's work is indispensable for understanding the disciplinary logic of the penitentiary system and the neoliberal logic of the control prison, it does not quite capture the *feeling* of living death described by prisoners in their experiences in carceral spaces, especially that of solitary confinement (Guenther 2013: xvii; see Chapter 5).

We must also consider the fact that incarceration – the abject conditions of confinement themselves – *breeds* violence; once incarcerated, formerly non-violent offenders often become violent ones. More than 75 percent of federal prisoners, for example, are incarcerated for a nonviolent drug-related offense, many for transporting drugs. Yet most of the prisoners confined to the most intense maximum security conditions in federal prisons are there not for a crime committed on the outside, but rather for some infraction that occurred *during* incarceration (Richards 2008; Haney 2008; Morris 2000). The reality is that prisons socialize prisoners, guards, and the entire prison administrative apparatus to be more violent. As I have shown in the context of the U.S. federal penitentiary system, the reproduction of conditions that provoke violence within prisons itself ensures longer prison sentences for the accused, 70 percent of whom are people of color (Morin 2013). The racial disparities of those condemned to death row or who have been executed are likewise striking: Black people comprise 34.7 percent of those executed in the U.S. since 1976, and 42 percent of those on death row, yet comprise only 13.2 percent of the U.S. population (Death Penalty Information Center 2017).

Thus the degree to which we can speak of the culpability, innocence, or victimhood of prisoners and animals that are exploited, abused, or killed within the Prison-, Agricultural-, or Medical Industrial Complexes is not as straightforward as it might first appear. If carceral logics target Black bodies for particularly easeful exploitation, as the biomedical research atrocities that Washington (2006) and Hornblum (1998; 2007) outline at Holmesburg Prison and elsewhere, the structural association of those bodies with those of nonhumans exploited in the lab are more obvious. To more fully develop the basis of this carceral logic I turn in the next section to relationships across 'animality' and 'humanness' which themselves create the conditions within which certain human and nonhuman bodies can be made exploitable and disposable.

Animality: human and other-than-human

Which humans and nonhumans do 'we' care about, and which do we not? Which humans and nonhumans have the force of legal, political, ethical, or other protections and which do not? Fundamental to how and why animals and prisoners can be exploited, objectified, and made killable within the research lab are the basic distinctions made between 'the human' and 'the animal'. These attributions are applied to various bodies via a number of different mechanisms and processes that either amplify the status of certain humans and nonhumans – 'humanizing' them – or that reduce the status of others and 'animalize' them. Processes of animalization are common throughout any number of social sites and institutions, yet they have particular salience within carceral spaces. I want to pause to explore these processes in some detail, as they supply a powerful driver of the carceral logics that determine which bodies can be made disposable and killable in the carceral spaces of the prison and research laboratory.

Manifestly, what may be done to or with nonhuman animals that represent various social categories – pets, performers, wildlife, livestock, vermin, beasts of burden, seeing-eye dogs, and so on (Morgan and Cole 2011) – mirrors the ways in which different human groups are endowed with humanness or animality and are subsequently protected or deprived of protections. Most vulnerable are 'animalized' lives such as vermin, pests, and livestock, and for my purposes here, racialized/criminalized Black and other minoritized people, who can be made exploitable as research subjects because they lack ostensibly 'human' qualities.

To put a finer point on this, both humans and nonhumans may be assigned 'human' or 'animal' characteristics. More than two decades ago Wolfe and Elmer (1995) offered a useful typology of how the human and the animal as ideological constructs intersect within various social spheres in hierarchical and value-laden ways. Their four types included: (1) "animalized animals" (these are who we eat, who we wear, who we test our products on); (2) "humanized animals" (these are pets and other animals endowed with putatively positive anthropomorphic/human features; (3) "animalized humans" (human beings subjected to all manner of brutalizations carried out by cultural prescription, as well as those who serve as reminders of human beings' bodily, organic existence); and (4) "humanized humans" (the category of pure human, "sovereign and untroubled").

These distinctions work to subjugate both certain humans and certain nonhumans into hierarchies of worthiness and value, distinctions that are highly calculated to reinforce human superiority. Indeed, it is important to recognize that using the terms 'inhumane', 'dehumanization', or 'subhuman' to describe particular types of conditions or subjects is problematic because such terms simultaneously place the human above and superior to the animal. In this vein Steiner (2005) traces the history and the logic of human supremacy over animals from pre-Socratic thinkers to the present, in which I would highlight the Judeo-Christian ordering of man's dominion over animals and the natural world as outlined in the *Book of Genesis*. The social meanings of both human and animal attach to various bodies and populations, and are then used to perpetuate hierarchical human–human as well as human–nonhuman relationships based on any number of social markers, perhaps most importantly racial ones. With respect to social meanings attached to 'the animal', as Kim (2010: 63) elucidates,

> In Kantian terms, animals are not rational, morally autonomous beings, which means they have no intrinsic moral worth. In thinking this way, we have thoroughly animalized animals – as slave, body, matter, nature, and object. Indeed, humans can denigrate other human beings through animalization only because animalness, though constructed, is such a stable site of meaning for us, an enduring counterpoint to humanness, the baseline

below which we cannot fall. As Keith Thomas (1983: 41) notes, "It was as a comment on *human* nature that the concept of animality was devised."

Dozens of scholars, including Patterson (2002: 27–50), Kim (2010; 2011; 2015: 24–60), Spiegel (1997), Socha (2013), Nast (2015), Glick (2013), and Hart (2014), have offered useful historical-geographical illustrations of the many human groups that have been vilified as animal or 'subhuman other', marking countless numbers of dominations, exploitations, and oppressions. Animalization of various human groups has played a central role in racializations, enslavements, genocides, colonizations, and imperialisms across centuries and continents. Patterson (2002: 27–50) exhaustively catalogs the ways in which human groups throughout history and societies across the globe have vilified other human groups as beasts, brutes, and apes, to pigs, rats, and vermin. Humans across space and time have been held captive, displayed, and made into animal spectacles. For example, it was common in 19th-century American (and European) geographical circles for returning explorers to publicly parade indigenous peoples captured from the Arctic or Africa as living, subhuman 'discoveries' within the empire-building enterprise, as I have written about in other contexts (Morin 2011). Newkirk (2015) and Carvajal (2014) discuss the "human zoo" in the U.S.: a Congolese man displayed in 1906 at the Bronx Zoo monkey house. A visitor there could see Ota Benga, a member of the Batwa people, displayed in a cage with an orangutan. The sign above the cage listed Benga's age, height, and weight. It also read: "Brought from the Kasai River, Congo Free State, South Central Africa by Dr. Samuel P. Verner. Exhibited each afternoon during September."

Without rehearsing the vast number or extent of such examples here, a brief overview of American wars and interventions abroad in the last century reveals dominant cultural representations of Philippino 'yellow monkeys' (1898), Vietnamese 'termites' (1969), and Iraqi 'cockroaches' (1991), to name just a few.[4] Within the U.S., African-Americans, Native-Americans, and Chinese-Americans have been targeted as the most animal of humans by the dominant culture – base, lowly, brutish, irrational, vicious, dirty, or lustful – so as to justify their subordination, exploitation, and extermination (Kim 2011: 329). Kim (2015: 24–60) argues that such associations first converged in a powerful way ca. 1860 when the Black, Native, and Chinese 'questions' arose in U.S. national consciousness. Within various contexts these groups came to occupy a marginalized borderland between human and animal. Their uses of and relationships to nonhuman animals (for example in San Francisco Chinatown's live animal markets) amplified accusations of cruelty and barbarism of these groups and corresponding racism, nativism, and cultural imperialist responses.

Deckha (2013b: 516–519) has shown that the 'discourse of civilization' itself permeates anti-cruelty legislation, targeting practices of minoritized groups' behavior toward animals as deviant or transgressive and thereby reinforcing race, class, religious, and gender hierarchies (also see Elder et al. 1998). Such legislation targets individual animal abusers who, through gross neglect, do not maintain adequate shelter, food, or veterinary care for animals. Meanwhile, industrial practices that abuse animals on a whole other (massive) scale remain immune from prosecution, irrespective of their violence against animals (Deckha 2013b: 526–526; see Chapter 4). This has the double effect of both selecting certain animals for non-protection ('animalized animals') as well as creating a deviant class of 'animalized humans'. Derrida's (2002: 394–395) formulation that "we kill animals instead of humans" is important within this context; the discourse of animality becomes so crucial because we indeed do kill humans. Therefore "noncriminal putting to death" (the denial of 'murder') within the slaughterhouse, the research laboratory, and the prison execution chamber must supplant the carceral logic of killing for both animalized humans and nonhumans.

Meanwhile, who or what is the 'ideal' human, that being who is exceptional and pure and against which other humans and nonhumans can be measured, will inevitably fall short, and will ultimately serve as a justification for violence enacted upon them? Sylvia Wynter's (2003) insights surrounding the ontological status of 'the human' have been influential on this point. She asserts that the Western or European, White, male, and bourgeois (and presumably heterosexual and able-bodied) human represents himself and has been overrepresented as the complete human, as if he were the only model of humanness, the apex, the epitome, the very definition and normative measure of the human. Thus there are "humans," and then all of "humans' Others" (Jackson 2013: 66). Jackson (2015: 215–216) adds that it has become common to encounter appeals to move beyond "the human" in diverse scholarly domains, but asks, *whose* conception of humanity are we moving beyond? To Ko (2016: n.p.), those belonging to this putatively ideal configuration of the human can afford to 'give up their humanism' within critical animal liberation movements because their humanity is already presumed, whereas for Others it is not. As Ko explains,

> that is why those of us who reside in, think, speak, theorize and exist on what I call the border of 'the human' and 'the animal' play a special role when it comes to the situation of animals ... we are all props for a narrative about "the human," a small group of people that are not only homo sapiens but an ideal genre of homo sapiens – which is why we are all kindred spirits in a fight to depose "the human."

Jackson (2015: 215–216) wonders if the call to 'move beyond the human' is in fact a call to 'move beyond race'. Within these discussions and relationships, 'Blackness' has assumed a particular salience. Kim (2010: 66; 2015), Nast (2015), Glick (2013), Sexton (2010) and others have argued that Blackness and animalness (along with criminality) are uniquely and mutually constituted and reinforcing, which further helps inform our understanding of the integration of 'animals' into carceral systems and spaces.

Blackness: criminal/animal

The over-determination of the Black body as criminal feeds a powerful carceral logic and process that creates and reifies the human–animal distinction. Black men are among the most abjected and marginalized of human groups, treated as 'not quite human humans'. Racial difference is basic to much of the 'criminal as animal' rhetoric, and animalistic representations of Black men in particular originate in various social arenas. As Wacquant (2001: 104; 2009) has argued, the wide diffusion of bestial metaphors for criminals in the journalistic and political field – "where mentions of 'superpredators', 'wolf-packs', and 'animals' are commonplace" – in turn supplies a powerful common-sense racist explanation for the massive over-incarceration of Black men, "using color as a proxy for dangerousness." Cacho (2014) in fact shows that certain activities cast as 'criminal' are almost unrecognizable without the Black body. In one example she offers, the same act of 'looting' a store is cast by the news media as unlawful when performed by Black people, but considered "resourceful" when enacted by Whites. Kim (2016: 42–43) adds that we have not only come to believe that all Blacks are violent, but that *"violence itself is black"*:

> so much so that when private citizens and state officials execute black men, women, and children for mundane actions such as sitting a car with friends, playing with a toy gun in the park, knocking on a person's door for help, and walking on the street, this is taken to be "reasonable" rather than "excessive" force.

Many prisoners themselves turn to animal imagery to express the shame and anger at being caged and treated like animals in a zoo (Morin 2015: 75; Hornblum 2007). Leder with Greco (2014) offer a first-hand account of the extent to which prisoners are likened to animals – they are savage, bestial predators who live in the jungle or zoo of the prison. Haney (2008: 963–969) describes prisoners reacting to their potential loss of humanity by becoming belligerent, spending their time pacing back and forth in their cells – "an image that is hauntingly similar to caged felines at the zoo." Many commentators submit that this association of Blackness with criminality is founded on a notion of the prison itself manifesting as the afterlife of the plantation, which is itself the afterlife of slavery – the prison as its latest form (Franklin 1989; James 2005; Gilmore 2007; McKittrick 2011; DuVernay 2016). (I return to this subject, particularly via a discussion of prisoner labor via the convict lease system, in the next chapter.)

Stepping back, Sexton (2010), Wilderson (2010), and Dayan (2011) provocatively posit that the ontological status of "Blackness" itself originated through Middle Passage chattel slavery, and out of this process, a foundational link formed between the African-American Black body and animality. While various human groups have been subjected to enslavement, extermination, genocide and other atrocities, they have not been excluded from 'the human' in the ways that enslaved Africans and their U.S. descendants have been. To Wilderson (2010: 1–52, quote on 38), something ontologically unique took place on the Middle Passage slave ship, a racialized transformation that cannot be undone: "Jews went into Auschwitz and came out Jews; Africans went into the [slave] ships and came out as Blacks. The former is a Human holocaust; the latter is a Human *and* a metaphysical holocaust." Racialized slavery wrought an ontological rupture that produced The Black as the very antithesis of a human subject – "the counterpoint against which the Human could gain coherence and knowledge of self" (Wilderson 2010: 9). Sexton (2010) argues that chattel (propertied) slavery displaced the human with the animal, representing a "tear" in the quality of being itself. Chattel slavery fused the biopolitics of the slave with the biopolitics of the animal, enabled by the fact that the biological Black *body* (versus the mind of reason) is itself overdetermined by visibility (after Fanon 1967). As Dayan (2011: 116) argues:

> What does it mean in times of torture and dissembling to *be like an animal*? It all began with chattels. Their treatment helps us to understand the limits of cruelty. They are used as examples when humans need most to categorize, to dominate, to justify slavery, genocide, and incarceration.

Chattel slavery was different from other forms of slavery in that it was racialized, which is why, to these authors, it would be problematical to assert that animals or anyone else for that matter, are today's "slaves" (e.g. Wise's 2002 and 2016 assertion that "all animals are slaves today"; see next chapter). Racialized slavery, then, created a unique association of Blackness with animality; and to be Black is to share in this ontological status (which is not, however, applicable to any individual Black person, but rather to the ontological status of Blackness, which has a tendency to erase difference). As Kim (2016: 45) articulates, the "human" then is paradigmatically "*both not-animal and not-black.*"

Whether or not one subscribes to Wilderson's (2010) and Sexton's (2010) arguments, they do help frame the 'oppression Olympics' arguments many have put forth asserting the relative horror of one atrocity or genocide over another in human history, and even while staying attuned to variances in scale versus kind (Spiegel 1997; Socha 2013; Patterson 2002). Many challenge the notion that the experiences of animals could be comparable to any human atrocity (e.g. Hart 2014: 673–674; Coetzee 1999). In this outlook, making such a comparison is to desecrate the memories of human lives lost and the 'sacrosanct moral divide between humans and animals', in effect reinscribing the very association of these groups to 'animalness' that justified their genocidal past in the first place (Kim 2011: 313, 326–330; Hart 2014: 673–677). (Similar arguments appear in works that animalize differently challenged or disabled human bodies; see Taylor 2013.)

The dangerous effect or alternative, however, is the re-subordination of the status of animals; the 'animality of animals' (Kim 2010: 68). Insisting that animals share human characteristics is similarly unproductive, though – i.e. in arguments over animal personhood (Wise 2002; Mowe 2016; Safina 2016) – since these simply reinforce the superiority of the putatively human and those animals who share their (our) characteristics. Again, identifying the processes by which certain humans and certain nonhumans are animalized does not require us to assert that they are 'the same', nor that such configurations are in any way stable and fixed over time, space, or cultural context – they are not. It is to say rather that animalization as a process subjugates and makes disposable – within the research lab and elsewhere – both certain humans and certain nonhumans and is rooted in deeply uneven hierarchies of power that provides certain groups with rights and resources and denies them to others (Lopez and Gillespie 2015: 3–7). Quite obviously, it provides the carceral logic that allows Holmesburg prisoners (Hornblum 1998) and DeMello's (2014) rabbits alike to become disposable material within the scientific research apparatus.

Animals such as mice and other rodents, while not criminalized, are 'always already' animalized and thus become acceptable research test subjects. Human prisoners become viable test subjects through a process that is entangled deeply with processes of racialization, criminalization, and animalization. Yet 'non-Black' prisoners may also be exploited as research test subjects. This tells us that criminalization and animalization are intertwined with a varied landscape of embodiment in carceral spaces that includes racialized Blackness but also includes other prisoners of color who are (also) disproportionately incarcerated in U.S. prisons (Native American, Latinx), as well as transgendered populations, and of course, the poor. Many people of color are permanently criminalized, and to use Cacho's (2014) phrase, are thus "ineligible for personhood" for different reasons – undocumented immigrants, the racialized poor from the global South, as well as criminalized U.S. residents of color from both inner city and rural areas.

As Kim (2010: 64–70) admonishes, attempting to compete for primacy or foundationalism of one form of oppression over another will always be problematical since oppressions are mutually constituted and reinforced; that the fates of various humans and nonhumans are inextricably linked – and certainly the carceral logic that allows certain human and nonhuman bodies to be exploited as testing material in the research lab provides a powerful example. As she states, we need to find "less taxonomic, hierarchical ways of relating" to others, and one way forward is in developing a 'multi-optic' vision that allows us to simultaneously view the 'optic' of animal oppression and struggle alongside the 'optic' of racial oppression and struggle (Kim 2015: 21–30, 181–182). We must challenge classifications, categorizations, and hierarchies among races and species, which will allow us to challenge the very boundaries between 'the human' and 'the animal' themselves. As Sexton (2016) and Ko (2016 above) argue as well, both human and animal liberation depend not on reifying but on "deposing the human"; i.e. that the logic of humanness itself can be 'deposed' by examining the logic that renders certain human groups 'animal'.

How the distinction between the human and the animal is made and unmade through carceral processes is a key feature of the above discussion.

Perhaps it is more obvious the ways in which animality is "made" through carceral processes, but what about animality being "unmade" through them? Thomas and Shields (2012: 6) argue that even though popular culture is fraught with prisoner–animal metaphors, the lack of species construction analysis in current prison studies is surprising. Thus, if we do not connect racialized animalization of prisoners to speciesism, we cannot fully address the problem of the prison (also see Socha 2013: 234–235). Deckha helps us see that 'race thinking' and 'species thinking' reinforce one another and work in combination. To Deckha (2010: 44–45), human rights discourses that attempt to purge the subhuman category, that argue that we should see all human beings as human, ultimately do not effectively achieve the aims of protecting vulnerable human groups from violence because they leave both the 'subhuman' and 'humanized human' categories intact; the binary holds.

> This viewpoint assumes that the impediment to humanism is its incomplete application rather than some defect in the category itself. Ultimately, instead of fighting dehumanization with humanization, a better strategy would be to minimize the human/nonhuman boundary altogether (Deckha 2010: 47).

The various laboratories wherein research on animals and prisoners takes place are different kinds of carceral spaces, each governed by their own carceral logic. Yet their processes are inter-related and deeply entangled. As Haraway (2008: 336N23) observed in a different context, it is not that various human or animal atrocities such as these have no relation, it is that "analogy culminating in equation can blunt our alertness to irreducible difference and multiplicity and demands. Different atrocities deserve their own languages, even if there are no words for what they do." Thus, rather than to merely argue that the speciesism that allows for testing on animals is like racism or sexism or other -isms, it is more productive to see that their carceral logics of domination are intertwined, especially through that of Blackness and animalness. These -isms are not homologous, but the ways that they deploy these carceral logics are deeply entangled with one another.

Conclusion: beyond the law, animals and prisoners

Describing her work as a research scientist, a biochemist colleague of mine confessed that she "could be a Nazi." In her lab experiments she injects rats with stress hormones, kills ("euthanizes") them, and then dissects their brains for analysis. Reflecting on a decades-long career involved in this kind of research, my colleague confided that sometimes she was unable to bring herself to the lab. "I had to compartmentalize my mind so that there was literally no connection between sentient beings and my experiments. At that point it was to follow the protocol and complete the task. I dreaded it" (Anonymous 2017).

This 'distancing', desensitization, and disassociation from the cruelty and violence taking place within my colleague's research lab could just as easily apply to the slaughterhouse, the execution chamber, and indeed, the concentration camp or numerous other historical or contemporary carceral settings. Lagerwey (2003) and Benedict (2003), for example, discuss the reflections of nurses on their roles in the Nazi "Euthanasia Camps" during the Nuremberg trials. They describe these nurses' 'compartmentalization and desensitization' of what was occurring around them, as well as the coping mechanism that most employed – that of considering their work as "caring" for the doomed camp victims. Such elisions might be necessary for maintaining their sanity, but as Tuan (1999: 116) has argued, such compartmentalization and disassociation should receive greater critical attention: "We need to be aware of, and reflect upon, these disconnections of ordinary life, for, unless we do, we risk slipping, when circumstances permit, into disassociative monstrosities" (also see Groling 2014). Tuan (1999: 110) argues that while love and affection seem to structure our relationships with our pets, we perpetuate everyday normalized violence against animals, inflicting pain unthinking.

I return to the ethical dimensions of these 'disassociations' and distancing effects in the Afterword. Many ethical and bioethical questions must be addressed with respect to laboratory experimentation on nonhuman animals (Greenhough and Roe 2011), including consideration of the types of relationships that develop between researchers and their nonhuman experimental subjects. By way of a conclusion to this chapter, though, I wish to briefly touch on some additional issues relating to the respective legal statutes and protections available to prisoners and animals within the carceral settings of the laboratory experiment.

Generally speaking, the legal mechanisms and oversight of biomedical research on prisoners relates most recently to the 2006 Institute of Medicine (IOM)'s commissioned report, *Ethical Considerations for Research Involving Prisoners* (mentioned above). This report advocated the simplifying of then-current federal regulations, and was codified legislatively as 45 C.F.R. 46, Subpart C (Research and Prisoners) (Reiter 2009). Prior to this report and the ensuing legislation, research on prisoners was restricted to four basic categories: (1) studies involving causes and effects of incarceration; (2) prisons as institutional structures; (3) conditions affecting prisoners as a class; and (4) research on practices that have the intent and reasonable probability of improving the wellbeing of the subject. The overall effect of the 2006 change was in loosening the protections that these categories represented and that had been in place since the 1970s (Hornblum 1998: xiv; Reiter 2009: 520–544).

The IOM report advocated a case-by-case 'cost-benefit analysis' of any research, and perhaps most problematically, recommended that the definition of "prisoner" be expanded to include not just those incarcerated in prisons, but also those released to probation and parole, or confined in non-traditional facilities such as drug treatment centers. This latter change added millions of potential test subjects annually, in effect creating the "largest population of controlled and traceable research subjects in the world" (Dober 2008), since parolees are required to notify state officials if they move, and are monitored by supervising officers on a regular basis. This shift returns us to the very norms challenged in the congressional hearings of the 1970s when Dr. Albert Sabin, who had tested his live virus polio vaccine on prisoners in Ohio, argued that prisoners were ideal subjects for research because they were "a stable, long-time permanent study group" (Mitford 1973). Ultimately, the legal revisions mostly demonstrate the powerful lobby of Big Pharma (Zellhoefer 2013: 253). Current regulations over research using prisoners are overseen by the Office of Human Research Protections (OHRP), a small unit within the Department of Health and Human Services. Notably, the OHRP regulations only apply to research that receives federal funding; there is effectively no outside oversight of privately funded research.

Today, those who oppose using nonhuman animals (especially primates) in research arguably comprise a more influential government lobby and have more public sympathy than those who oppose using prisoners for research. Moreover, research on (some) animals is regulated by a more complex array of overlapping regulations than that for human prisoners (but recall here too that the 'animals' protected do not include most nonhumans used in testing, such as rodents and rabbits). To take one example, the National Research Council's *Guide for the Care and Use of Laboratory Animals*, funded by the NIH and other government agencies, is 120 pages in length. But the *Office of Research Protections, IRB Guidebook*, devotes only four pages to using prisoners as research subjects.

The (Laboratory) Animal Welfare Act of 1966, amended many times and later becoming known as the AWA, is the main federal law regulating the treatment of animals in laboratory research. The Act requires minimum standards for "handling, housing, feeding, watering, sanitation, shelter, care, and psychological wellbeing" of animals. The Act requires that animals receive a balanced diet of wholesome foods, environmental enrichment (see Fig. 3.3), a physical environment adequate to promote the psychological wellbeing, and clean and structural housing with suggested dimensions. (Of course, whether the regulations on care of animals are followed is another matter.) The 1985 Improved Standards for Laboratory Animals Act also required the establishment of IACUCs (Institutional Animal Care and Use Committees), a parallel to the IRBs for research involving vulnerable human subjects. Importantly, and again, the AWA only mandates these requirements for warm-blooded animals – it does not protect reptiles, birds, rats, mice, or farm animals (Braverman 2013a) – certainly by the numbers the very populations of nonhumans who would need protections in laboratory research settings. And importantly, as with the oversight of research involving prisoners, the oversight of laboratory research involving animals is required for federally funded research only; no such outside oversight exists for private facilities. Moreover, there is little or no record keeping in federal agencies, for example 'no pain and distress reports' required (Gruen 2011).

As Gruen (2011: 126–128) has argued, the best way to see what is wrong with experimentation on animals is to compare cases where humans have been reduced to laboratory material and the 'affront to humanity' that such work has represented. She points out that at least with the case of (some) humans, protections are in place to avoid coercion, require informed consent, and minimize risks; whereas plainly speaking, animals cannot give informed consent (above). Thus there is "an obvious human exceptionalism at work here." Gruen (2016b) has painstakingly identified by name the First 100 chimpanzees who were subjected to biomedical research, and is in the process of documenting the names of the Last 1,000 who are being moved from laboratory to sanctuary, as a way of "honoring their service":

> For almost 100 years, chimpanzees have been used in biomedical and behavioral research in this country, the last industrialized country to experiment on our next of kin. The end of using chimpanzees as nameless test subjects is near. Already hundreds of research chimpanzees have been retired. In tribute to all who have been forced to serve, here we look forward to the journey to sanctuary of the Last 1000.

It might be worth asking whether we could imagine such a 'naming' project of the thousands of U.S. prisoners who were forced to serve and lost their lives or their health for science. As I argued above and as Gruen (2011) would undoubtedly agree, the notion of 'informed consent' of prisoners is wholly fraught within medical, emotional, and economical parameters. Heading down this slippery slope of human exceptionalism, though, will be unproductive when the vulnerable humans in question are prisoners. One could observe for instance that the suggested enclosure dimensions within the AWA for two 66-pound gorillas or chimpanzees represent more square footage than the U.S. Department of Justice (via American Corrections) allows for two human prisoners who share a typical prison cell in today's maximum security prison (Morin 2013). Thus rather than competing for worst-case oppression, employing the 'multi-optic' perspective that Kim (2015) advocates would offer an opportunity to see both the norms of these respective captivities, and the scientific research undertaken within them, as wholly unacceptable.

Notes

1 One of the results of the Nuremberg trials of Nazi war criminals was the development of the 1947 International Nuremberg Code, which, with numerous aspects, focused on the ethical principles and legal capacity of individuals to consent to becoming test material.
2 A number of important topics that could lend themselves to discussion in this chapter are not directly addressed; for example the gendered nature of the research lab. Most research subjects, both human and nonhuman, are male, and some practitioners have brought needed attention to what will be missing in clinical trials via this bias (e.g. Bichell 2016).

3 As Gruen (2011: 107) observes, more and more research today involves 'chimeras' – a single organism comprised of cells from different individuals, as in a stem cell transplant. As the PCRM (2017) states, there is no way to know how many mice, rats, fish, or birds are used in research because these animals are excluded from federal laws that require scientists to report their numbers. Estimates are, though, that approximately 95 percent of animals in U.S. laboratories are mice and rats.

4 Many scholars, including Philo (1998), have moreover shown that humans working and living in the spatial borderlands between humans and animals, such as those living and working close to quarters of livestock destined for slaughter, are themselves cast as bestial.

4 Laboring prisoners, laboring animals

Introduction

The play *La Historia de Nuestras Vidas* (*The Story of Our Lives*) describes the personal experiences of 400 undocumented workers who were arrested in the 2008 raid on the AgriProcessors Meatpacking Plant in Postville, Iowa (Skitolsky 2008; Raymond 2009). These workers wrote *La Historia* to describe the ICE (Immigration and Customs Enforcement) raid on the plant, their arrest, their transport to a nearby fairground, the subsequent legal charges leveled against them for 'aggravated identity theft', and months of imprisonment at various detention centers across the United States. One section of the play describes the workers' experiences of being held in a former site of cattle slaughter, and their newly raised consciousness about the treatment of animals at the site:

> They had prepared a place to hold us ... a place meant for cows.
> Like the cows we butchered at AgriProcessors.
> But now we were the ones being processed.
> Now we were the ones being "processed."
> They made us remove our pants and shirts
> They took all of our clothes, everything we carried.
> I was shivering from the cold.
> I was trembling with fear.
> They took us from our rooms, one or two at a time ....
> We were all loaded into the back of a moving truck – herded together with chains on our feet, like pure animals.
> Like cows.

The play speaks poignantly to the similar torturous treatment enacted upon these workers and the cattle they processed every day at AgriProcessors. Yet the dialogue also hints at an underlying implication that these human workers – though having developed a new affinity and perhaps empathy for those animal lives – deserved to be treated better than mere cows. Many scholars and culture critics have observed the vulnerable, poorly compensated, at-risk, and often undocumented workers exploited in the animal processing and meatpacking industries, and due to these workers' close proximity to animals, have noted the cultural associations made to their 'subhuman' animal status (e.g. Franju 1949; Philo 1998). Significant scholarly attention has been paid generally to the exploitation, job insecurity, abysmal working conditions, and overall precariousness of human lives, especially of immigrants, within the agri-processing industries (Schlosser 2001; Patterson 2002: 61–63; Loyd et al. 2012; Moran et al. 2013; Sittig 2016; Grabell 2017). Many cite Upton Sinclair's influential muckraking novel *The Jungle* (1906), which explored early 20th-century U.S. slaughterhouses through the eyes of the immigrant worker. Yet the 'hierarchies of worthiness' embedded in valuing workers' lives over those of animals in many of these studies prevents us from adopting a 'multi-optic vision' necessary to grasp how such carceral spaces and industrial regimes of power, profit, and greed 'require' the respective oppressions of humans and nonhumans alike (Kim 2015: 181–182).

In this chapter I focus on these respective oppressions and exploitations and their accompanying carceral logics, taking inspiration from scholars who have studied the similarities in accumulation strategies, commodifications of life, and devaluations of labor across human prisoner and numerous nonhuman groups. Various human and animal bodies are routinely and similarly marked, managed, commodified, and consumed based on capitalism's drive for profit (Spiegel 1997; Harvey 1998; Hudson 2011; Nibert 2002; 2013; Hart 2014; Gillespie and Lopez 2015). I focus on the labor of people ensnared within U.S. carceral institutions, historically and to the present, and nonhumans laboring within various spaces of captivity. I examine what their positioning in U.S. labor and economic relations can tell us about industrial violence, their lives being made into property and commodity, and the legal, political, and economic contexts that make this possible. This chapter extends my discussion of "commodifying vulnerable populations" in Chapter 3 in important ways. What legal frameworks governing property status, property relations, commodity production, and commodity relations allow prisoner and animal lives to be made exploitable and disposable? And what formations or expressions of state sovereignty allow their lives to be made ultimately killable? In the service of capital, science, and putatively 'safe streets', the Prison-, Medical-, and Agricultural Industrial Complexes collectively determine whose lives can be exploited as profit for others.

Ava DuVernay's (2016) documentary film *13th* has drawn widespread popular attention to the exploitation and commodification of prisoners historically and continuing today in a juridically-sanctioned, modern form. The title *13th* refers to the 13th Amendment to the U.S. Constitution, passed by Congress in 1865, which abolished slavery *except* as punishment for a crime. As my discussion below outlines and DuVernay's film showcases, the roots of modern prison labor can thus be found in the rebuilding of the U.S. South during the 19th-century Reconstruction, when millions of Black people were criminalized for petty offenses, imprisoned, and put to work – an economic explanation for the mass incarceration and exploitation of labor we see continuing today. As DuVernay describes the situation today:

The [case] that was most disconcerting to me was Victoria's Secret. They [used prisoner labor] until it was made public, and then changed their practices, and so did J.C. Penny. Weapons systems, pieces of missile defense … household goods are made by people who are forced to do it because they are imprisoned, and they get paid pennies. These are practices that we would not allow in sweatshops, yet [are] O.K. for folks serving time (Buckley 2016).

As I discuss in detail below, the Prison Industrial Complex (PIC) itself comprises a collaboration of many intertwined interests of government and industry, one of which is most obviously that large-scale imprisonment ensures one type of solution or spatial 'fix' to local or regional economic shortfalls by supplying a captive workforce for the production of a wide array of goods on the inside (Gilmore 2007; Bonds 2009; Palaez 2014). The private prison and its ancillary private companies such as Corrections Corporation of America (CCA) also today earn unprecedented profits on the bodies of the incarcerated through control of individual sectors of the PIC such as in health care and food service delivery, and most recently, in handling the 'business' side of probation and parole.

Nibert (2002; 2003; 2013) was among the first animal studies scholars to argue that the shared devaluation, economic exploitation, oppression, and vulnerabilities of numerous human and nonhuman groups are inextricably linked to one another – in his case, arguing that speciesism is structurally similar to racism, sexism, and other human oppressions. Nibert found both theoretical and empirical resonances across oppressions of racism, sexism, and speciesism and the historical roots of oppression in the devaluations of each's labor. He offers a useful starting point in thinking about shared oppressions, though Nibert's critics note that his work does not account for cruelty and exploitation of laboring animals in pre-market societies or under other (non-capitalist) economic systems; nor that such exploitation and cruelty can proceed from other human self-interests beyond or outside of the economic; nor that market 'compassion' can be sold to consumers just like other products (Jasper 2003; also see my discussion in chapter 3 of the problems associated with equating these '-isms'). Employing a strictly economic-system rationale to argue that a change to a different (non-capitalist) mode of production would in some way guarantee animal protections will not likely hold; if consumers want cruelty-free products, companies will appear to provide them and make a profit doing so. Thus in the following discussion I attempt to consider primarily various legal but also political and cultural lenses and norms that work through a capitalist economy to enable the exploitation of prisoner and animal labor, and through which resistance to these norms could take shape.

In the subsequent discussion I engage with scholars who have based their arguments about the economic exploitation of captive humans and nonhumans on their respective status as property and/or commodity (Francione 2004; 2008; Wadiwel 2015; Collard 2014; Shukin 2009; Deckha 2013a). One might consider as a starting point to such a discussion asking whether those held captive might 'own' their own bodies within a carceral site, and if so, in what sense or to what extent? And with such ownership, what legal protections or rights are available to those made to work in carceral settings? Similarities as well as differences can be observed across the Prison-, Agricultural-, and Medical Industrial Complexes with respect to the status and rights afforded prisoners and animals as 'persons' (or the lack thereof), and the treatment that then can follow. With respect to animals, most prominently Wise (2000) has argued that certain animals such as apes, elephants, and dolphins ought to be considered persons by virtue of their sameness with humans (shared cognitive abilities, emotional capacity, etc.). These are 'honourary humans' (Deckha 2013a: 805). As I discuss in more detail below, the most obvious limitation to this line of thinking is that it reinforces the animality, the lack of personhood – and thus the status as property – of the vast majority of other animals who do not share those 'human' characteristics and whose bodies can be made disposable and their labors exploitable.

Billions of animals, by virtue of their status as property, are brought into existence simply to be slaughtered, to labor for humans, to be enjoyed as human entertainment, and to be used as human surrogates in biomedical research; and yet the rules that govern these practices are limited to only where and how they are raised, sold, and killed, even in comparison to treatment of 'wild' animals (Wolfson and Sullivan 2004; Braverman 2013b: 121). Becoming clothing, meat, or other commodity is at least partly dependent upon becoming property first. These animals all perform 'work', but they exist in various capacities vis-à-vis labor relations and production; and they derive their values respectively as 'live' property or commodities, or as dead ones. That said, an animal does not have to be considered property in order to be subjected to violence or bodily appropriation, nor does an animal body need to be considered property in order to be commodified (e.g. consider the African safari). Property, then, does not explain all conditions under which animals suffer and are commodified (Gillespie 2012: 121; Collard and Dempsey 2013).

The legal status of animals and (some) prisoners within carceral spaces makes it easier to commodify their labor or their bodies, but this is not the only condition of their commodification. Marginalized and warehoused prisoners – otherwise structurally irrelevant to the market (Peck 2003) – also lack many rights of 'personhood' which allows for exploitation of their labor on the inside. Many argue that prisoners exploited as such retain their status as chattels through the ongoing slavery of the prison 'plantation' (DuVernay 2016; Franklin 1989; James 2005; Gilmore 2007; McKittrick 2011). Yet if the prisoner is considered enslaved property, who 'owns' the prisoner? The state? What legal frameworks and relationships allow prisoner labor to be derived out of property relations and/or commodity relations? Is the exploitation of prisoner labor sustained more via 'commodification' than that of property? And if so, what legal, political, or economic mechanisms sustain these distinct modes of commodification of prison labor?

As I wrestle with these and related questions below I conclude by briefly examining some ways that the process of labor exploitation inexorably transforms prisoner and animal bodies in carceral spaces. Some sort of transformation necessarily takes place within these carceral settings, a transformation from which the subject can never return (Shabazz 2015b; McKee 2015; Collard 2012).

Laboring prisoners

Franklin (1989), James (2005), Gilmore (2007), DuVernay (2016) and numerous others have argued that the U.S. penitentiary can be considered no less than the latest version of the slave plantation. Some prisons, such as the Louisiana State Penitentiary in Angola, are literally built on former plantations; and people of color, although comprising only 30 percent of the total U.S. population, comprise 60 percent of those imprisoned. Today, one in three Black men can expect to be imprisoned at some point in their life (compared to one in six Latino men, and one in 17 White; Alexander 2012; ACLU 2017). The prison as an important site of captive, exploited human labor finds historical roots first in racial slavery (Chapter 3) and subsequently in the convict lease system that lasted 50 years following the emancipation of slaves, and today continues to find expression with millions of U.S. prisoners working on the 'inside' at numerous prison industries.

Oshinsky (1996), Cabana (1996), and Gilmore (2007) offer comprehensive historical views of the use of prisoner labor within the convict lease system. After the Civil War and the constitutional abolition of slavery, a host of ostensibly legal maneuvers were enacted designed to guarantee the availability of cheap Southern Black people's labor; laws that allowed for the imprisonment of men for long periods of time for minor crimes – for example, outlawing both 'moving around' and 'standing still'. Gottschalk (2006: 49) argues that the defining feature of post-emancipation Southern politics was the leasing of convict labor for half a century after the Civil War. Despite being promoted by local, regional, and national governments as a progressive and reformist response to the problem of "Negro crime," conditions were in many ways more brutal than on the plantation, since there were no incentives to invest in the well-being of a leased convict.

Both Oshinsky (1996: 139–155) and Cabana (1996: 111–139) focus on the Parchman Farm – what is today's Mississippi State Prison – a huge 20,000-acre penal farm that resembled its predecessor, the antebellum plantation designed to control and exploit Black labor.[1] By 1915, Parchman was a self-sufficient operation containing a sawmill, a brickyard, a slaughterhouse, a vegetable canning plant, and two cotton gins, while pouring almost a million dollars into the Mississippi state treasury through the sale of cotton and cottonseed. Parchman resembled an antebellum plantation – both the plantation and the prison used captive labor to grow the same crops in identical ways. With a dependable supply of convict workers, the prison farm produced some of the highest quality cotton in the entire Delta from the work of hundreds of men who spent long, back-breaking hours in the fields. Parchman remained an intensive work farm into the 1980s. Both the plantation and the prison also relied on a small staff of rural, lower-class Whites to supervise the Black laborers, and both staffs mixed physical punishment with paternalistic rewards to 'motivate' their workers. A similar scenario played out at Louisiana's State Prison at Angola. Prison agriculture, embedded within historical legacies of plantation slavery, colonization, and hierarchies of power, remain clearly visible to this day (Gillespie forthcoming).

Under the convict lease system, state authorities and prison officials also hired out prisoners to private contractors who exploited their labor in coal mines, turpentine farms, sawmills, phosphate pits, and brickyards. Officially banning the use of convict labor was one of the professed reforms of the 1920s Progressive Era, and yet the reforms only allowed the state to more directly exploit prisoners, putting them to work building roads, constructing railroads, and developing the infrastructure and economy of the South using the abominable 'chain gang' (Gottschalk 2006: 49–51; Gilmore 2007). Chain gangs became notorious for horrendous conditions; prisoners labored in fields six days a week, 10 hours a day, with other prisoners brutally serving as 'trustees' guarding them and punishing them with guns and whips (see Fig. 4.1). Chain gangs served an integral role in the federal "good roads" movement of the mid-20th century, not dismantled

Figure 4.1 Chain gang of juvenile convicts in the U.S., 1903. Library of Congress, Prints and Photographs Division, under the Digital ID det. 4a28370.

until the Civil Rights era – and then, only to be revived three decades later (Lichtenstein 1996; Myers 1998). In 1995 the chain gang reappeared as a method of punishment in Alabama and Arizona prisons, "thus resurrecting a powerful and shameful symbol of this country's legacy of racial injustice and institutionalized racial injustice" (Gorman 1997; see Fig. 4.2). The chain gangs under infamous Maricopa County (Arizona) Sheriff Joe Arpaio are but one of many 'law enforcement' human rights abuses that took place under his recent and long watch (Jerreat 2014). (Arpaio was finally ousted in the November 2016 elections and subsequently convicted of wrongdoing, only to be pardoned in a highly contested political move by the U.S. president the following August.)

'Employing' prisoners to labor on the inside, along with vocational education and recreation, were the hallmarks of the Progressive Era prison reform movement. 'Reform over retribution' of the burgeoning prison population served as the underlying logic of the creation of the Federal Bureau of Prisons itself in 1930, based on the idea that work (and play) were necessary for rehabilitation. Many federal and state prisons were located and built in rural areas where prisoners could work on prison farms or in the on-site factories. In the case of the local penitentiary where I live in Pennsylvania (USP-Lewisburg), prisoners farmed poultry, dairy cattle, hay, corn, clover, soybeans, alfalfa, sorghum and potatoes, and others worked in the prison's metal factory (Morin 2013: 385).

Figure 4.2 Modern chain gang. Photograph courtesy of Patrick Denker, 2006.

So an important question to consider in the context of today's penal philosophy – which has clearly moved in the direction of warehousing and punishing offenders rather than reforming or rehabilitating them – an important question is how labor (paid but also unpaid) within prisons is understood within the corrections industry (and hence, by the U.S. Department of Justice). Do there remain lingering notions of work as rehabilitation within some corners of corrections, and/or is prisoner labor simply to be understood as one manifestation of the capitalist-incentivized privatization of the prison industry, in whole or in part? One way to begin answering that question is to simply ask, who benefits from prisoner labor?

Since the mid-1990s, the AFL-CIO reported that tens of thousands of prisoners working at below market wages were generating more than a billion in sales (Gottschalk 2006: 30). Today one of the most lucrative areas for the corrections industry is in contracting out prison labor at sub-minimum wages to private firms. McCormack (2012) and Pelaez (2014) argue that one of the explanations for the growth in prisons over the past four decades is that long prison sentences maximize the use of cheap prisoner labor and also populate private prisons both as property and as cheap labor sources, even as crime rates go down. As Pelaez (2014) argues,

> the private contracting of prisoners for work fosters incentives to lock people up. Prisons depend on this income. Corporate stockholders who make money off prisoners' work lobby for longer sentences, in order to expand their workforce. The system feeds off itself.

Prisoners in the U.S. today make office furniture, fabricate body armor, and manufacture textiles, shoes, and clothing (among others) – while typically earning tremendously deflated wages of between .93 cents and $4.73 per day, the profits of which accrue to the state or private enterprises. Thirty-seven states have legalized the contracting of prison labor by private corporations that have established their operations inside state prisons; these include IBM, Boeing, Motorola, Microsoft, AT&T, Dell, Nordstrom's, Revlon, Macy's, Pierre Cardin, Target, and many others (Pelaez 2014; DuVernay 2016). Some prisoners in state penitentiaries may receive a minimum wage for their labor but this varies tremendously by state; in Colorado they receive $2.00 per hour, and in private prisons as little as .17 cents per hour.

DuVernay (2016) and Elk and Sloane (2011) alert us to the influence of the lobbying consortium ALEC (American Legislative Exchange Council). ALEC is a group representing corporations such as Walmart, State Farm, Big Pharma, AT&T, and Verizon that exerts powerful legislative influence to maintain the conditions under which prisoner labor can be exploited – including shaping crime policy and advocating for longer prison sentences (e.g. SB1070). Corrections Corporation of America (CCA, above), part of the ALEC group, holds a federal contract to hold 'illegal-looking' detainees, with 11 million people currently detained in their facilities each month. As alluded to previously, to protect against potential revenue loss if the sentencing laws become relaxed in the coming years, ALEC and the likes of the CCA are preparing the Prison

Industrial Complex for alternative revenue streams, such as in privatizing probation and parole via the production of GPS monitoring devices, community supervision, and other surveillance technologies. As DuVernay (2016) astutely observes, "we should all be suspicious when all of a sudden the ALEC group supports crime reform," because their underlying motivation will always be in anticipating new business opportunities.

Blinder (2015) reports that the Mississippi state system of exploiting prison labor is currently under scrutiny; it relies on prisoner labor to launder jail uniforms, repair government vehicles, collect roadside litter, and clean high school athletic facilities and government buildings, saving the state tens of millions of dollars each year that it would otherwise have to pay in regular worker salaries and benefits. The state had guaranteed during the 1990s jail building spree that they would never be less than 80 percent occupied, and so they boosted local revenues by sending state prisoners to local jails, paying the per diem via taxpayer dollars while receiving free labor from prisoners. This made it easier for local governments to absorb budget cuts in state funding. Now, however, with a potential decline in the prison population, officials worry about how harmful locking fewer people up will be for the local economy, with income deficits and budget shortfalls without free prisoner labor. "Since some counties still owe significant sums on bonds that financed the jails, they are now considering bringing inmates from out of state to cover the shortfall" (Grim 2016a; also see Grim 2016b; I return to this below).

Returning to the question of who benefits from prisoner labor, it is obvious that the close collaboration between government and industry protects and maintains a number of corporate business interests and profits. One should be skeptical of arguments made by corrections personnel that such work primarily aids prisoner rehabilitation or reform, or what 'rehabilitation' itself could even mean within today's punitive era of mass incarceration (e.g. Phelps 2011). But despite the conditions and the pay, a similar argument about the 'volunteerism' needed for medical testing on prisoners (Chapter 3) could be made with respect to prison labor. That is, that prisoners 'want' to work because it provides a break from the monotony of prison life, gives them a place to be for hours each day, and offers more money than they would otherwise have. Moreover, the correctional industry's justification for work programs rests on contentions that they develop vocational skills and accrue other social benefits to prisoners (see below). Crete (2017) offers an interesting Canadian First Nations example of the ostensible 'healing' that takes place within prisoners by their producing Native artifacts on the inside. Yet clearly, carceral settings offer inherently narrow parameters within which such 'choices' could be made, and which are ultimately for the instrumentalist and profit-seeking benefits of others.

Laboring animals

> Milk Is Our Life's Work.
>> Captioned 'cow-speak' appearing on trucks and marketing materials,
>> Upstate Niagara Cooperative.

In contrast to prisoner labor, when it comes to an analysis of 'laboring' or 'working' animals we confront a wide range of activities in form and type, as well as a diffuse array of contexts and spaces within which captive animals labor for the benefit and profit of humans.[2] My attention is drawn primarily to animals laboring at sites within the Prison-, Agricultural-, and Medical Industrial Complexes but clearly animal labor extends well beyond these, for example throughout the 'Entertainment Industrial Complex'. Perhaps what this spatial ubiquity alerts us to more than anything else is the profound fluidity by which human carceral logic has been and can be deployed in putting animal bodies to work.

First, it is not only human but also nonhuman labor that can be exploited within U.S. prison walls. Perhaps the most well-known example would be the animals made to perform with human prisoners at rodeos in Oklahoma and Louisiana. The Louisiana State Prison at Angola is the largest maximum-security prison in the country; it covers 18,000 acres, almost the size of Manhattan, and holds over 5,000 prisoners in sprawling housing units. In addition to the expansive farm where crops are hand-cultivated by prison workers, Angola also features a popular (prisoner-maintained) public golf course on prison grounds, as well as a rodeo stadium where five or six times a year prisoners perform stunts for thousands of spectators with cows, bulls, horses, and other captive animals (Schrift 2004; Turner 2013; Gillespie forthcoming). Such 'dark tourism' capitalizes on the public's fascination with criminality through the spectacle of live 'animalistic' – and untrained – prisoners competing in events with their 'wild' animal counterparts. Such human and nonhuman labors serve as both entertainment and a voyeuristic spectacle of punishment.

Dogs especially have a long history of working in both military prisons and in regular corrections (Urbanik 2012; Haraway 2008). At Mississippi's Parchman and other turn-of-the-century prisons (above), dogs were used to hunt runaway prisoners. Today, therapy dog training has become a common practice in many prisons; the aim is to ostensibly reform prisoners while preparing dogs for occupations on the outside, especially as companion animals. Haraway (2008: 63–65) documents programs such as those at Pelican Bay State Prison in California, in which prisoners are taught how to train their canine cell-mates who later become workers on the outside, including as police attack dogs.

Such (literal) incarceration of dogs and other animals requires careful scrutiny. Moran (2015b: 11; who also makes the distinction between prison 'pets and pests') has argued that we should pay careful attention to the nature of prison confinement across species since a dog's (or other animal's) experience in prison may be one of 'loving companionship'. She suggests that "prison dogs may experience prison programming in much the same way as domestic dogs experience being family pets, benefitting from human company and engaging in the kind of kinaesthetic empathy widely observed amongst companion dogs." McKee (2015) posits a similar example of prisoners and ex-racehorses and their respective potential as 'redemptive capital' (below). Urbanik (2012: 85) also discusses 'service animals' and therapy dogs and horses, confined in prisons or at many other sites, what she labels a "modern twist on working animals." They are trained to "socialize, work gently, remain calm, complete tasks, and stay focused." They and other animals in the military and law enforcement – such as pigeons, horses, and elephants – have served as couriers, as guards, as test subjects, and as transport. Today, 2,700 dogs serve the U.S. military alone, in everything from bomb detection to helping improve troop morale (Urbanik 2012: 75).

Prisons, military zones, and other sites of law enforcement and policing are but one type of governmental, industrial, or cultural site where captive animals perform labor for human benefit. Urbanik (2012: 75–102), Philo and Wilbert (2000), Haraway (2008), Hribal (2003), Clark (2014), Gillespie (2014) and many others have offered broad views of what we might think of as locations of 'laboring' animals – the multitude of spaces and places in which humans utilize and exploit nonhumans as workers. Animals have been launched into space. They appear in movies and are made to perform for human pleasure in racing, circuses, dog and rooster fighting, and rodeos. They labor for humans at zoos, water parks, and other spaces of 'caged' entertainment (Chapter 5). On a recent search of my local pet animal rescue facilities I found that the American Society for the Prevention of Cruelty to Animals (ASPCA) identifies some of its cats available for rescue as "working cats," meaning that they will eliminate rodents from infested areas. Animals, of course, also serve as educational tools and 'material' within a vast array of sites of clinical research and experimentation. Clark (2014: 139–140; after Hribal 2003 and Abadie 2010) offers a compelling case to consider animals – like humans – as 'laboring' at such sites; they are an integral part of the production cycle – with researchers they are 'co-workers' in the lab. Such "'guinea-pigging' ... is a particular kind of work; instead of doing things it mainly entails having things done to you." It is useful to acknowledge that animals' laboring in clinical or other research represents a type of vulnerability that they share with their incarcerated human counterparts (Chapter 3).

Suffice it to say, though, that the farm is probably the most ubiquitous site of animal labor and exploitation. For millennia across the globe animals have worked in agriculture for transport, ploughing, planting, hunting, and herding (Urbanik 2012; Emel et al. 2015; Hribal 2003; Nibert 2002; 2003). Arguably 'draft' animals used in agriculture have made up the largest category of working animals (Urbanik 2012: 83) – all of those horses, donkeys, sheep, mules, goats, dogs, and others used to plow fields, pump water, herd other animals, and transport people and goods. As such, Hribal (2003) recognizes animals as historically "part of the working class." Tuan (1999: 114) observes that dogs have been bred, historically, for use (rather than pleasure). To be useful in hunting, they must be obedient. He notes that nearly all the small dogs that we now think of as playthings – terriers, spaniels, and even poodles – were once bred for hunting. In many places animals continue to do farm labor; they are literally physical laborers – such as horses pulling tractors and other farm equipment through Amish and Mennonite fields in Pennsylvania and Indiana.

Defining animal 'work' or labor on the farm or in the agricultural sector more generally varies over time, place, and culture. Guenther (2013) and Hudson (2011), among scores of other scholars and commentators, argue that the modern farm is an institution of carceral confinement comparable to the prison, and within that institution we find animal 'labor' both *producing* or being *produced as* commodities (see Fig. 4.3). The distinction of farm animals who are generally relegated as 'working subjects' as opposed to 'worked objects' is an important one. Animal bodies are transformed in their billions into meat, other foods, clothing, and numerous other commodities within the factory farm system (the CAFO – concentrated animal feeding operation). We might imagine such animals – like those guinea pigs who labor in the research lab, who passively 'have things done to them' – as, alternatively, both instruments and objects of the production cycle; as co-workers in the production cycle; or as one among many actants in a 'vast distributive and heterogeneous assemblage' that is the meat-producing sector (cf. Clark 2014: 160).

The potential 'agency' of animals on the farm or in the research lab is crucial to this distinction, yet it is just as crucial with respect to *any* captive animal labor on the farm (cf. Emel et al. 2015). Clearly, animals routinely serve as laborers in

Figure 4.3 Prisoners milk cows confined at the Colorado Correctional Industries (CCI) facility. Photograph by Colorado Correctional Industries.

Figure 4.4 Chickens confined to battery cages laboring in egg production. Photograph
courtesy of Creative Commons Share-Alike.

producing other commodities – such as chickens producing eggs; pigs, horses,
and cows 'contributing' extracted semen for forced (re)production of others; and
dairy cows producing milk (see Fig. 4.4). Collard and Gillespie (2015: 1) evocatively describe the end-of-life scene of one such laborer in the dairy industry:

> The cow lies on the manure-covered floor of a holding pen at the dairy
> market sale ... Her past as a typical dairy producer is easily visible on her

body in the scars on her hide, the blood and milk leaking from her udders, her docked tail and ear tags, and her emaciated appearance. A sticker with a barcode is stuck haphazardly to her side with #743 … She lies on her stomach, her back legs splayed out behind her, unable to rise. She struggles several times to get to her feet and each time collapses with exhaustion, breathing heavily, saliva foaming at her mouth … Because she cannot stand, she will be categorized as a "downer' and shot with a firearm at the end of the day. A 'deadstock' hauler will be called to pick up her body and deliver her to a rendering facility.

Thus I would argue that an expansive definition of what we might consider animal labor or animal work is useful and necessary. Haraway (2008: 73) urges us to "take animals seriously as workers":

we might nurture responsibility with and for other animals better by plumbing the category of labor more than the category of rights, with [the latter's] inevitable preoccupation with similarity, analogy, calculation, and honorary membership in the expanded abstraction of the Human.

Finally, it is important to observe with Gillespie (2012: 118) that dairy farm animals are "subjected to routine forms of bodily modifications and appropriation that signal their status as property: branding, ear-tagging, tail-docking, nose-ringing, castration without anesthetic, and artificial insemination" – the range of things that can be done to a farm animal who is considered the property of another. The issue of animals as property of humans has been a major concern in a wide range of contexts, and it is to that issue I next turn.

Animals: property and commodity

Many scholars have argued – Francione (1995; 2004; 2008) perhaps the most forcefully (but also see Dayan 2011; Braverman 2015; Wadiwel 2015; Deckha 2013a) – that the ability of humans to exploit nonhuman bodies and labor relies fundamentally on the legal status of animals as human property. Depending on how one counts them, most animals can become human property; they are commodities that those with human status can own, and as property owners, humans give them value. Animals are owned 'just like cars or furniture' (Francione 2004: 116–117); and as their possessions humans can use them, gain from them, sue for their fair market value, employ them as collateral, and routinely kill them.

Creatures constituted as property would seem to be without rights, and yet as I have touched upon in previous chapters, some animals do have partial 'rights' bestowed by the state via protective laws and legislation. Dogs are an interesting case; the law recognizes dogs as both property and as partial persons. Anti-cruelty legislation demands reparation for harm in the name of the dog's sentience and suffering, as well as the harm done to the dog owner's property.

Pet dogs are also considered property when they can be neutered, bred, and euthanized, all with deference to the owners' objectives. As Srinivasan (2013: 109–110) reminds us, "it is the very idea of property" that allows for the categorization of 'stray' dog. In comparison to dog as property in the UK, Srinivasan observes that the law in India recognizes the independent status of ownerless street dogs – the concept of a dog that cannot be owned, and moreover, cannot be killed. Yildirum (2017) warns us though that such nomadic dogs may nonetheless be subjected to the same detention and policing strategies as those of human refugees; in her case study, via relegation to 'street dog reservoirs' in Istanbul. When dogs are used for something else, though (e.g. food), they cease to become dogs altogether (Srinivasan 2013; Haraway 2008; Dayan 2016).

Even though the creation of commodities out of animal bodies is enabled at least in part through their status as property – in the food system, in experimentation, in fashion, and in entertainment – not all nonhuman animals are constituted as property in the way or to the extent that farm animals are; 'wild' or free-roaming animals are not considered property in the same way, for example (Braverman 2013b: 121; Deckha 2013a).[3] Nor does an animal need to be considered property in order to be made into a commodity or to be subjected to violence or bodily appropriation – consider for example hunted or safari animals. Status as property, then, does not explain all conditions under which animals suffer and are commodified. And yet, those nonhuman animals cast as having more relative 'animality' – pests, vermin, and livestock – are considerably more vulnerable to violence and killability within processes of labor, production, and consumption than are those considered as having more 'human' qualities (Wrye 2015; Kim 2015; Gillespie and Collard 2015; Clark 2014).

Wadiwel (2015: 147, 150–157, after Francione 1995: 35) locates the continual appropriation of animals as both property and commodity as nothing less than an "act and artefact of war." He focuses on the violence of jurisprudence, which always balances the property status of animals against unchallenged and sovereign human interests. Yet, he observes, "an animal is different from inanimate property, such as a tool," and thus the law makes the distinction that animal property not be "wasted or that animals not be killed or made to suffer when there is no legitimate economic purpose." Ignoring the ambiguity of what might be considered 'legitimate' suffering, we can note that in food production, for example, animal slaughter is a value-creating act; the death is timed to maximize the relative exchange value of the meat that will be produced. As such, as Wadiwel (2015: 163) shows,

> death is not threatened to the animal as an incentive to labor (as in slavery), since death will be imposed at the time that is deemed to be most profitable … at this point value is realized through a forced exchange of life for death. … [H]ere we strike a point of exchange in production, which essentially describes the realization of the value in the commodity.

Wadiwel goes on to describe this process as "absurd ... [since] there can be no equivalence in value between life and death."

Relevant to my purposes here, Castree (2003: 275–276) advocates that geographers pay closer attention to what they (we) mean by the 'capitalist commodification of nature' and the processes by which it takes place: 'what is the nature of the materials being commodified and the circumstances of their production?' He defines a commodity simply as "something that can be sold and/or exchanged," while "commodification ... describes the process by which previously non-saleable and non-exchangeable things become commodities" (Castree 2004: 25). Collard and Dempsey (2013: 2684) helpfully outline some variables that matter in the determination of which animals live and which die in the process of their commodification. Certain animals derive their value through their death (as meat, as lab material), whereas the 'liveliness' of others is extracted for profit (circus animals, trafficked pets). Animal labor commodified in the wildlife trade and in the entertainment industries, for example, relies not on animal deaths but on their liveliness, wildness, and 'encounterability'. A particular mode of value-generating life predominates in these commodity circuits. Collard (2014: 153), for example, refers to trafficked pets such as parrots and lizards as "lively commodities" – commodities whose active demonstrations of being full of life – their labor – is more important to their value than even their sentience. As an intervention in Marx's labor theory of value Haraway (2008: 47) adds that these captive "undead things" then provide a third form of value alongside that of use value and exchange value: that of trans-species "encounter value."

As these examples amply illustrate, a fundamental component of our capitalist economy is based on animals' commodification and value even if they are devalued politically and ethically (Collard and Dempsey 2013: 2691). Shukin (2009: 6–20) adds a cultural dimension to an understanding of these commodity circuits by alerting us to the "profound contingency" of market life on animals as both "figures" and flesh. She employs the double meaning of "rendering" of animal life to link together the (physical) industrial boiling down and recycling of an animal body and the reproducing, interpreting, or representing animals in linguistic or other artistic means. Her phrase, "animal capital" then, refers to the ways in which animals are made usable as both material ('carnal') objects as well as usable as cultural artefacts (an automobile advertisement serving as just one example). This happens through the "simultaneous capitalization of nature and naturalization of capital."

Legal geographies of animal property and commodity

> Law is anthropocentric terrain.
> Maneesha Deckha (2013a: 783).

Although focusing on the law will only offer a partial picture of what we might more generally consider animal (or, for that matter, human) 'rights', it is nonetheless important to recognize the state's legal protections – or lack of them – that govern the exploitability, commodification, and killability of nonhuman animals. Essentially we can observe a limited number of laws that relate to a limited number of practices that involve a limited number of non-human animals within a limited number of sites and institutions. The legal system deems certain animals as exploitable and killable, and others as protected and un-killable. The essential question to ask is, 'what status must one have in the eyes of the law in order to be exploitable, disposable, and killable?' As Taylor (2013: 540–541; after Wadiwel 2009: 285–287) explains, the exercise of law is the means by which domination of certain humans and nonhumans is 'encoded, naturalized, and continues, emanating from the right to death held by the sovereign'. It is obviously fully legal to routinely kill certain animals; and it is animals within the agricultural industry who are lacking the most in legal protections. Moreover, quite obviously, animal death never enters into the "realm of murder" in the eyes of the law, so as Glick (2013: 645; after Mbembe 2003) argues, "animal death already belongs to the realm of the "necropolitical reign of terror," with everyday life marked by the sign of con-stant death; it is banal, routinized, "everywhere and nowhere."

Braverman (2013b: 105–108) points to the importance of the on-going legal classification of animals – whether wildlife, pests, pets, companion animals, or livestock – with respect to their protection or exploitation. For animals to be regulated, they first must be "defined and sorted," either through a Linnean classification system or through their relationships to humans – an ever-shifting process that produces an 'inherent messiness and fluidity to animal regulations' (also see Wolch and Emel 1998: xvi). Thus as Braverman continues (2013b: 108–111), the vast majority of animals fall into one of the following "legal orders": wild, domestic, agricultural, pests, and laboratory animals – all of which are not internally consistent categories nor are they mutually exclusive. Historically those most protected were either considered part of nature (wildlife, especially those endangered or threatened), or were part of the human community (pets); those not protected were constituted as sources of profit and value.

As Francione (2004: 114) explains, there are two kinds of criminal laws in the U.S. that govern humane treatment of animals, 'general' and 'specific'. Anti-cruelty laws are general, ostensibly referring to all animals, whether owned as property or 'un-owned'. Specific laws govern use of animals for specific uses, such as the Animal Welfare Act of 1966 that governs experimentation on (some) animals, and the Humane Slaughter Act of 1958, which governs the treatment of animals produced as food (and despite excluding most nonhuman animals used as food, such as chickens, turkeys, other birds, as well as rabbits and fish). (Cf. Elder et al. 1998 and Deckha 2013b on how these very anti-cruelty laws 'sustain a discourse of civilization' and reinforce dominant racial, religious, classed, and gendered hierarchies by targeting minoritized people and practices as deviant or transgressive.) However, a much greater number of laws and legal frameworks enable and protect the rights of those who use animals as testing material, food, and laborers than those that protect the animals.

Various legislative acts make it a felony to gain access to and record what takes place in slaughterhouses (Pachirat 2011: 5–7). Powerful 'Big Ag' lobbies such as those that represent the food producer Aramark ostensibly argue for the protection and better treatment of farm animals but in actuality attempt to ensure the availability of meat. The powerful lobby of 'Big Pharma' works for similar ends with respect to animals used in research (Zellhoefer 2013: 253). As of 2012 the legislatures of ten states had passed "ag-gag" laws aimed at criminalizing the taking of photos or videos of animal cruelty and abuse in factory farming. 'Animal facility fraud' has become a new criminal category, argued on the grounds of protecting private property (Kyle and Sewell 2015; Rasmussen 2015). (Zellhoefer 2013 also offers a useful historical analysis of the Animal Enterprise Protection Act of the 1990s, and the subsequent and more draconian 2006 Animal Enterprise Terrorism Act, aimed at silencing animal rights activists.) As Patterson (2002: 71) asserts, "The American meat and dairy industries have successfully convinced their friends in state legislatures and Congress that what agribusiness does to animals should be 'beyond the law'."

However, one important intervention in this legal maneuvering occurred in Idaho in 2012. A video showed workers at Bettencourt Dairies in Hansen, Idaho, stomping, beating, and abusing cows, including using a tractor to drag a cow on the floor by a chain attached to her neck. Despite the Idaho ag-gag law, a federal judge ruled the repression of such a taping to be a violation of the right to free speech. This was the first time an ag-gag law was declared to be unconstitutional. As the *New York Times* editorial board wrote,

> While most Americans enjoy eating meat, it is hard to stomach the often sadistic treatment of factory-farmed cows, pigs and chickens. Farm operators know this, and they go to great lengths to hide these gruesome images from the public. These ag-gag laws purport to be about the protection of private property, but they are nothing more than government-sanctioned censorship of a matter of public interest. The judge in the case was quoted as saying that 'food production is not a private matter' (Exposing Abuse 2015; also see Bittman 2011).

As Rasmussen (2015: 64) further argues, "the agricultural spaces regulated by ag-gag laws are not best understood as conflict between human and animal interests but [rather] … by the demands of biocapital that establishes relationships of power between some bodies and others."

While house pets and other animals are protected via anti-cruelty laws (although they themselves are time-, place-, and context-specific; Elder et al. 1998), most farmed animals are exempt from them, illustrating again the contradictory legal universe that is 'animal protection'. As Francione repeatedly observes, humans suffer a collective "moral schizophrenia" about their animal property (2004: 108; 2008: 26–28): what we say about the importance of moral and ethical treatment of animals stands in stark contrast to how we actually treat them. Moreover, the law is extraordinarily vague on what "cruelty" means

(Wolfson and Sullivan 2004: 211; Deckha 2013a: 789–790). Because cruelty refers to activities that are ostensibly 'unnecessary or unjustifiable', anti-cruelty laws exempt suffering caused in food production – thus the throwing of a live lobster into a pot of boiling water is wholly justifiable, the pain and suffering irrelevant. Yet it is essential to not ignore what might be the subjective experience of cruelty, methodologically speaking (see Chapter 1).

Most animals are condemned to live without legal rights. Farm animals are exempt from protection because states' legislative statutes exclude or exempt 'common' or 'customary' farming practices from any oversight. As in the oversight of zoos, it is the zoo managers and industry itself that defines the criminality of its conduct – not prosecutors, judges, and juries (Donahue and Trump 2006). As Wolfson and Sullivan explain (2004: 206–217) "whether 'customary' farming practices are cruel or not we should be concerned that the farming industry has complete control over it." They note that 98 percent of egg-producing chickens live in battery cages, eight hens to one 20 by 19-inch cage. "A judge merely has to determine whether the practice is 'customary' to determine whether challenges could be brought." Such laws ought to concern us: 9.5 *billion* farm animals are slaughtered annually in the U.S., compared with a combined 218 million killed by hunters and trappers, in animal shelters, in biomedical research and dissection, and product testing. These numbers are possible because farm animals are fundamentally considered property, and farms are not typically inspected by regulatory agencies. Moreover, laws regulating the transportation of farm animals refer only to transport by rail, while most animals are overwhelmingly transported by truck (Wolfson and Sullivan 2004: 208). And even if a penalty were imposed for improper transport, the maximum penalty is a mere $500.

Scholars such as Wise (2000) have argued in the media and in the courts that changing the legal status of animals from property ("legal thinghood") to "persons" would secure the same legal rights that are available to humans. He argues that some animals – particularly chimps, apes, gorillas, dolphins, elephants, orcas, and parrots – should be provided the legal status of person due to their human-like higher order cognitive abilities and emotional intelligence (also see Wise 2016; Nonhuman Rights Project 2015). While Siebert (2014: 32) alerts us that normally one must have 'standing' in court to speak of one's own injury or to bring legal action; yet habeas corpus precedents have allowed others to represent those who are unable to personally appear in court. He notes though that a 'legal person' is not the same thing as a human being; that numerous entities have acquired legal status as person but are not human beings, such as corporations, and most recently, the Whanganui River in New Zealand. Thus, why should 'personhood' rights be reserved for humans only?

Francione's work (1995; 2004; 2008) disentangles whether the 'equal consideration' of human and animal sentience and suffering is a viable approach to protecting animals from harm. He observes the limitations of such reasoning, since the dividing line between humans and nonhumans will always be arbitrary: the 'abilities test' is arbitrary ("why is it better to be able to recognize oneself in the mirror than it is to be able to fly or breathe underwater"?). Moreover, and more problematically, equal consideration arguments simply reinforce the 'superior' human standard against which other nonhumans would be compared and judged as lacking. And indeed, many humans do not possess 'human' characteristics, yet we would not consider them property. Francione (2004: 123–129) likewise argues for the legal personhood of animals, but key to his argument is in aligning the property status of animals to that of the historical status of slaves; i.e. that human slavery was (is) structurally identical to the institution of animal ownership, and it was not until the slave was granted the status of a legal person rather than property that human labor could no longer be owned by another. To Francione (2004: 131), all sentient beings have morally significant interests, and should have the right to not be treated as the property of others.

The bifurcation between animal property and personhood is central to most animal rights debates, a debate that has reached something of a stalemate. Braverman (2015: 307–309) argues that we need a new way of thinking about law and society when it comes to animals, moving beyond the 'personhood' threshold of Western law altogether, towards what she calls a 'more-than-human-legalities' discourse. "Both animality and humanness are deeply embedded in the construction of the law … and law is acutely relevant for constituting the animal." To her, jurisprudence should be opened to the foundational questions of the meaning of governing and being governed, and the ethical and political questions that emerge in governing not only human but also more-than-human life. Deckha (2013a: 790–792) offers a useful example of what such a non-anthropocentric opening, a 'relational' valuation of animals in the law, would look like.

Within U.S. family law companion animals have been treated as more than ordinary property, as family members. Some courts have applied the 'best interest' test in deciding with whom a companion animal should reside and whether one spouse owes the other 'petimony' when the spouses separate; and other courts have included companion animals in estate planning as trust beneficiaries rather than as property. In addition, 19 U.S. states have changed terminology concerning the relationship of companion animals to their humans from 'owner' to 'guardian' (Deckha 2013a: 790–792). As significant as these developments are, as Deckha points out, they are not really about respect for the animal qua animal, and do not include the interests of most animals who do not reside in close physical or relational proximity to humans (and/or who are exploited for food, experimentation, or entertainment). In Canada, an opening of another sort occurred with a Supreme Court judge's dissent in the *Reece* decision that argued for the full "legal voice" and "legal subjectivity" of Lucy the zoo elephant and other animals based on their sentience, inability to consent to various practices, and vulnerability that is 'similar to human disadvantaged groups' – all of which fundamentally challenge their status as property (Deckha 2013a: 800–802). These non-anthropocentric openings in law allow for the possibility of representing animals in non-commodified ways, however much they continue to rely on human sentiment towards the animals for protections.

Again as we consider the legalities specifically governing carceral spaces, it is the carceral logic of the human–animal divide that guides reasoning, and the process of animalizing certain groups who are then subjugated and made more exploitable. As I continue to engage with that tension, I consider the limitations of considering prisoners as property or commodity, specifically the limitations of prisoners as 'legal personalities' who can be abused with impunity, bereft of legal rights and protections. What we see is a notable and important intersection between governmental and industrial biopower and various legal statuses, frameworks, and geographies that allow for the exploitation of their labor and/ or reinforcement of their status as property or commodity. As Gillespie and Lopez (2015: 6–7; after Foucault 2007: 144–145) assert, "this is not to imply a coherent unity ... of the state, but a 'governmentalization' of the state expressed as a multitude of forces that bear down on populations and is operationalized though economic logic."

Prisoners: property and commodity

As I have detailed in previous sections and chapters, many human as well as nonhuman bodies are similarly marked, managed, consumed, and killed by industrial capitalism's drive for profit. Hudson (2011) outlines the linkages across different types of 'surplus' populations, human and nonhuman, illustrating the capitalist means of making a profit on bodies that otherwise have no value and hence must be 'created' as a resource. Both human prisoners and nonhuman animals can become commodities and resources that can be bought, sold, and traded, although they of course play different roles in production. Guthman (2011: 237) offers hints at some of the ways that human bodies are endemic to capital flows 'far beyond their function as labour power ... the ways they form into property'. Her examples include bodies used for transportation of drugs; bodies mined for organs and sperm, eggs, and other genetic material; bodies rented out for human reproduction and other biomedical uses; and bodies bought and sold in various trades.

It is not only prisoners' labor that can be commodified, but their bodies and lives themselves can be bought and sold as well. With prisoners, in addition to laboring for abhorrently low wages on the inside, the profits of which accrue to the state and private entities (above); many local and regional economies depend on the income generated from the 'purchase' of incarcerated bodies from other jurisdictions to continue filling carceral sites that were built during the 1980s and 1990s construction boom (e.g. Grim 2016a; 2016b). Indeed, the carceral logics behind many prison-construction-as-economic-development strategies vary across those that suggest an ostensible current 'crises' caused by housing local prisoners elsewhere; to those that foresee future projected numbers of incarcerated bodies exceeding current capacity; to those that promise new revenue generated by housing prisoners from other jurisdictions. As I discussed in Chapter 3, the basic carceral logic and 'business' of the Prison Industrial Complex writ large – jobs, infrastructure, services – requires a reproduction ('breeding') of a steady supply of criminals in order to prevent from failing those local economies that are based on prisons and jails (Gilmore 2007; Peck 2003; Wacquant 2001; Wolfers et al. 2015; Bonds 2009; 2015). The carceral logics of 'profits through punishment' then, not only exploit prisoner bodies as things to be bought, sold, and traded; but also typically fail to provide promised community benefits such as job creation (Bonds 2015; Moran 2015a: 61–66; Norton 2015).

As noted above, a number of scholars equate the exploitation of animals as tantamount to their "enslavement"; that structurally, exploitation of animal bodies and labor is similar to that of human chattel slavery based on the status of both as the property of others (e.g. Spiegel 1997: 27–28; Best 2014). Hart (2014: 676–678), among others, raises objections to this association, for example in pointing to the different ways that enslaved humans versus nonhumans respond to the conditions of their enslavement. We need not rehash the 'oppression Olympics' discussed in Chapter 3 here; but I am persuaded by the many commentators who argue that the association of Blackness with criminality is founded on a spatial association of the prison itself manifesting as the afterlife of the slave plantation, the prison its latest form (Patterson 1982; Franklin 1989; Davis 2003; James 2006; Gilmore 2007; DuVernay 2016; McKittrick 2011). Although the specificities of prison life and plantation life differ greatly, as McKittrick (2011: 956) asserts, the plantation was nonetheless the "blueprint for the prison industrial complex, it anticipates and empirically maps it":

> the kind of labour performed while locked up, the "intent" of the containment, the racial-legal scripts of criminality, all lead to very different contexts through which articulations of violence and race take place. Yet the generalized traits of both institutions (displacement, surveillance, and enforced slow death) draw attention to the ongoing racialized workings of spatial violence.

Imprisonment in the U.S. today as a form of racialized slavery links to the obvious historical connections with the convict lease system following Reconstruction (above), the 13th Amendment that abolished slavery except as punishment for a crime (above), and particularly the obvious racism embedded in the judiciary and correctional system, evidenced by the over-representation of incarcerated minorities, especially Black people. As Gordon (2011: 11) emphatically asserts, "It remains the case that slavery is constitutionally enabled in the United States or any place subject to their jurisdiction for prisoners and for prisoners only." If the prison is more or less tantamount to the plantation, and the prisoner is tantamount to the slave, the prisoner also can be thought of as the de facto property of the state. Indeed, Gordon (2011: 16) calls the prisoner the "fungible property of the state." The extent to which prisoners officially enjoy the legal status or constitutional rights of 'citizenship' – such as voting rights, the right to free speech, freedom to practice religion, and so on, varies considerably by local, state, and federal jurisdictions, legal statutes, and cultures. It is not my intention to flesh out these differences here (particularly as it seems that prison administrators tend to follow their own rules on the inside anyway; Alexander 2012; James 2005; Morin 2013; see below). But again, the exceptional clause of the 13th Amendment allowing that prisoners can be treated as enslaved laborers on the inside points to them as a particular form of property and commodity from which to make a profit (also see Dayan 2001; 2011; Guenther 2013: 46–47; Cacho 2014).

The carceral logic of prisoners as animal – or prisoners as 'Blackened' which itself begets animalization – further enables their enslavement within the prison. Exploitation of labor (and killing itself) within the prison is enabled by animalizing the human and 'isolating the nonhuman within the human', since animal bodies can be exploited and killed without the commission of a crime. Humanness, then, is made a political, conceptual category rather than a biological fact, since certain humans can be defined as no longer fully human or deserving of human rights (Hudson 2011: 1664; Agamben 2004: 37; Dayan 2001; 2011; Glick 2013: 646; Deckha 2010). Via Wilderson (2010: 11–15) and Sexton (2010) I would add that through this process of stripping 'humanity' from the racialized slave/prisoner, and associating the slave/prisoner's Blackness with animality (and criminality), the route is even more easily paved for commodification of prisoners' bodies and labor: 'That which cannot be produced as a commodity, that is the Human'. To these authors, the material conditions of the slave/prisoner – which could be good or bad – are secondary to the political status of the slave/prisoner as property.

Moreover, as Cacho (2014: 6–7, 23–24) shows us in example after example, there is a *permanence* to the criminalization of people of color. 'Permanently criminalized' and racialized people do not need to break the law in order to be criminalized, and through this process they become "ineligible for personhood" – they are populations subjected to laws but refused the legal means to contest those laws and ultimately, subjected to social death. For different reasons, undocumented immigrants, the racialized poor of the global South, and criminalized U.S. residents of color in both inner city and rural areas are populations who "*never* achieve, in the eyes of others, the status of living" (Cacho 2014: 6–7). Tying this back to Lockian liberal philosophy, Cacho adds that the very idea of (White) personhood in U.S. history derived from the potential to own property: "The subject is defined by his ability to 'own' ... [and] the first and foremost thing he owns is himself."

It is within the carceral logic encircling the relationship between property and personhood that some animal rights critics argue that the only route to protecting animals from bodily and labor exploitation is to change their status from property to person, following the route that human slaves took from status as property to person (e.g. Francione 2004, above). Slaves escaped being considered property when they became legal 'persons'. And yet, as Dayan

(2011: 137) insightfully claims, unlike the animal, 'both the slave and free' were incorporated into the human slave body. She explains,

> the logic of slavery both depended on and tried to evade the consequences of the comparison with animals. Those who would make such a comparison had at once to explain how a slave could be property in the manner of animals, yet have a differential status. Pushed to give slaves some definition with the realm of reason and unreason, human and brute, southern jurists puzzled out how best to relegate a thinking agent to the status of property.

Dayan's insights suggest a consideration, then, of the prisoner as property without an owner; property and 'person' united in one body. Clearly a complex set of overlaying structures and legalities produce myriad forms of criminalization, enslavement, racialization, and commodification of bodies within the prison, and collaborations across government and industry, all occurring simultaneously. Wadiwel (2015) prefers to describe the slave as having a "commodity experience," while recognizing that slavery is also beyond mere commodification – the slave was property of a sort. The tensions presented by this 'property versus personhood' discussion, and the relationship of both to the sovereign state, prompts the question of how far this carceral logic can extend to the prisoner. In the case of animals, for Wadiwel (2015: 22–23), "sovereignty is anchored in human positionality, not through just a property right, but through an overarching system of domination that both encompasses and exceeds animals as property." A far-reaching set of relationships occurring between prisoners and the sovereign state, including but not exclusive of a 'property' relationship, articulates over and over again. It is to some of the political, legal, and economic contexts of these relationships to which I next turn.

Legal geographies of prisoner property and commodity

What is the legal status of prisoners that enables forms of violence to be enacted against them – the institutionalized carceral logic that allows exploitation of prisoner labor in the service of capital accumulation specifically? The question of who is exploitable (and disposable, killable) is intimately intertwined with the question of who or what has the power to exploit (and dispose, kill). What, then, we must ask, gives the U.S. sovereign state the legal (or other) right to treat prisoner bodies as property or commodity, or the right to enact any form of violence against them for that matter (Gillespie and Lopez 2015: 180)?

The statutory or 'official' government oversight of *federal* prisoner labor comes from the U.S. Department of Justice's Federal Bureau of Prisons and its "Federal Prison Industry" Program (FPI or UNICOR), a government-owned prison labor corporation in existence since 1934. Policies guiding the FPI are among the 338 policy documents within federal corrections itself, and within the FPI are 51 policies relating just to oversight of industrial production and labor (FPI 2017). These policies concern everything from accounting procedures to handling and dispensing of inventories, developing products and maintaining quality control, costing analysis, testing procedures, and marketing and sales. "Prison Made Products and Services" (2003) outlines which goods can be produced within federal prison walls, and "Work Programs for Inmates" details the purpose and scope of using prisoner labor within the Federal Prison Industry. As this latter document details,

> It is the policy of the Bureau of Prisons to provide work to all inmates (including inmates with a disability who, with or without reasonable accom-modations, can perform the essential tasks of the work assignment) confined in a federal institution. Federal Prison Industries, Inc. was established as a program to provide meaningful work for inmates. This work is designed to allow inmates the opportunity to acquire the knowledge, skills, and work habits which will be useful when released from the institution. There is no statutory requirement that inmates be paid for work in an industrial assignment ... [although] discretionary compensation to inmates [is provided by statute] (Kane 2017: 1).

In 2016, only 10 percent of eligible federal prisoners worked in a prison industry (17,900 people) – a tellingly small number compared to those whom the Bureau evidently considers a security risk and prevents from working, or for whom work is simply unavailable due to the gross expansion of the prison population. They earned .23 cents up to a maximum of $1.15 per hour (and by law, if they incurred any court-ordered financial obligations they were required to use half of their earnings towards those debts). (See National Correctional Industries Association 2015 for other programs and details.)

Individual state prisons and some local municipalities also have their own correctional industries, with their own work programs and guidelines to administer them. For example, in my state is the Department of Correction's 'Pennsylvania Correctional Industries' (PCI 2017), whose products, not inci-dentally, go by the brand moniker 'Big House Products'. By way of a quick comparison, in 2005 PCI employed 1,640 prison workers within its 18 state prisons, which represented 3.9 percent of the 41,000 prison population that year. They earned an average hourly rate of .59 cents and similar to the federal system, 30 percent of wages were required to pay various forms of "restitution." These Pennsylvania prisoners, according to the PCI website,

> produced nearly 1.7 million license plates, 46,000 pairs of inmate work boots, and nearly 12,000 pairs of eyeglasses; [they] canned almost 114,000 cases of fruit and vegetables, washed 13 million pounds of laundry, and processed 4.1 million pounds of beef, pork, turkey, and fish (Performance Audit 2005; Barnes 2005).

As in the federal system, these work programs are putatively intended to 'reduce inmate idleness, provide valuable inmate vocational training and work experience, and aid in re-entry and reduce recidivism'. Although the 2005 audit did find that the programs reduced prisoner idleness and taught vocational skills, the audit also reported many performance irregularities, including insider rebates, questionable accounting practices, a phenomenal waste of taxpayer dollars, and evidence that the programs had no impact whatsoever on prisoner recidivism despite aims to instill skills and a work ethic in prisoners.[4]

While the above provides a brief snapshot of official narratives behind prison work programs and their legal justifications – to provide (not necessarily paid) "meaningful" work and the "knowledge, skills, and work habits which will be useful when released," such a narrative does not begin to map on to what the exercise of sovereign power within the site of the prison actually looks like – nor does it take into account the often insurmountable social prejudices that prisoners face when attempting to be hired after serving time. Among staff and prisoners countless deal brokering, insidious manipulations, and terror-invoking atrocities of every sort and scale take place daily behind the closed doors and 'lawless' space of the prison, which I have examined in a number of other contexts (Morin 2013; 2016; Morin and Moran 2015). As I have shown specifically within the federal penitentiary system, judicial law and jurisprudence can be completely suspended in day-to-day prison operations, which often evidence capricious, masculinist exercise of power by guards, with administrative decisions – such as idiosyncratic systems of rewards and punishments – made internally by prison staff, not by judicial or court order.

Agamben (1998: 170; 2005) theorized that sovereign power can suspend the rule of law in such spaces by claiming "states of exception" in the midst of perceived crises or threats (e.g. the War on Drugs, the War on Terror). Within such a scenario, 'camp'-like spaces of exception are produced and contained within them are 'bare lives' ('homo sacer') – prisoners existing so far outside of the rubrics of citizenship and rights such that anything that happens to them is acceptable; no act committed against them could be considered a crime (Agamben 1998: 96, 71ff; Rhodes 2009; Czajka 2005: 130). Here is it useful to consider briefly the distinction between *social death* and *civil death*. Whereas social death means separation from family, kinship, and what makes life a liveable life (Butler 2009; Guenther 2013); as Patterson (1982) has argued, civil death is in relation to the state (i.e. statelessness) – it is to be 'dead in the law', to have no recourse to the law (Gordon 2011: 10; Cacho 2014). Prisoners are subjected to both. Their lives are reduced to exploitable biological existence, and stripped of political status and rights. One empirical measure of this power of the sovereign state is that lawsuits brought against prison bureaucracies typically have (only) about a 4 percent chance of success. As the American Civil Liberties Union reports, prisoner labor specifically has long existed as a "legal black hole" (Kovensky 2014).

Whether 'regular' correctional institutions – as opposed to death camps, Guantanamo Bay prison, or other U.S. detention centers or 'Black Sites' that Agamben studied – can be considered spaces of exception is debatable. And yet, it must also be noted that the suspension of prisoners' rights within regular correctional institutions is enacted as much *through* the law as outside of it; prisoners are captured within the political order but also represent the 'hidden' foundation of sovereign power that has been normalized (Moran 2015a: 19; Morin 2013: 388; Gregory 2006). As Hudson (2011: 1663) argues, sovereignty depends on the paradoxical location of the sovereign as both inside and outside the law, a paradoxical location which both produces 'others' and strips away their rights. For my purposes here, the legal but abhorrently low wages paid for prisoner labor on the inside, and the recent resurgence of the spectacle of the abominable chain gangs in Arizona and Alabama (above), offer no better examples of prisoners existing both inside and outside the law. Again as Cacho (2014) forcefully argues, the processes of criminalization of (already crim-inalized, racialized) people subjects them to the law's discipline without offering them the law's protections.

The relationship between bare life and political life through the exercise of sovereign power also offers important insights to distinctions made between humans and nonhuman animals. Agamben's (2004) work is full of interesting arguments about how the distinction between humans and animals was estab-lished in connection with one another, and how important that distinction is to modern notions of sovereign power, citizenship, and various legal statuses. Agamben himself (1998) relied on human life (only) as the basis for the concep-tions of 'killability'; to him, it is the juridical order that determines the 'human' through the capacity for politically meaningful deaths. Thus, reiterating the above discussion, 'humanness' itself is made a political, conceptual category rather than a biological fact – as is animality. In this way certain humans can be defined as not (or no longer) fully human or deserving of 'human rights'. As Deckha (2010: 38) articulates, it is the subhuman that is key to the carceral logic of the camps, that justifies the absence of the rule of law. Moreover, the human homo sacer who becomes politically disenfranchised becomes killable, but not 'murderable' in the legal sense. Animal death, likewise, does not enter into the realm of murder in the eyes of the law (Glick 2013: 645; Derrida 2002: 394–395); in many contexts and through the exercise of sovereign power, animals have no political agency, no rights, and can be killed with impunity. While all living things can be killed, only the 'human' can be murdered.[5]

The camp-like carceral spaces of the prison or the farm foundationally and in similar ways are able to turn any life into 'bare life'. As Hudson (2011: 1664) argues, "within these spaces, human beings are stripped of citizenship, denied a political voice … they are reduced to the animal. Spaces of exception are akin to the proliferating zoos, agro-industrial farms, and scientific labs where animals are kept." Moreover, it seems apparent that the sovereign power that renders both prisoners and animals killable within these spaces of exception also, via interlocking carceral logics, turns them into 'living property' or commodity. The usefulness in juxtaposing the carceral techniques in play with respect to U.S. prison labor programs with those of animals laboring as property or commodity on the farm allow us to see that it is particularly the linked carceral techniques of animalization (and racialization/criminalization) of certain populations, alongside a number of linked legal potholes and mechanisms, that create the conditions under which profiteering on prison and animal bodies takes place today at a demographic scale previously unknown.

Concluding reflections: agents of 'redemptive capital'

With respect to the complexity of life in spaces of exception – 'carceral camps' if you will – Agamben (1998; 2005) failed to reconcile or elides crucially important acts of resistance to bare life that take place within them. Foucault (1977: 308) offers a glimpse to the fact that while carceral power seeks to incapacitate, neutralize, and discipline the wild, such exercise of power is never complete. Factory farmed animals, for example, despite an extremely limited range of activities available to them, do resist being reduced to property or commodity; they bite, kick, produce less milk, and so on (Taylor 2013: 549). Wadiwel (2015: 167) observes that there are always such fractures to sovereign power:

> A chicken struggles against a human operator, as it is thrust into the poultry shackle; a hooked tuna fish, gasping, wrests its body violently on the deck of a ship; a cow hesitates before being prodded to enter the kill chute; a hog turns away as the captive bolt pistol misses. Animals intervene to prevent [an] exchange .... The will to prefer life over death is a primary act of resistance to the process of biopolitical commodification, perhaps the only act of resistance available to animals who are subject to extreme forms of control within production systems.

And because these moments of resistance "are seen as threats to the efficient accumulation of capital, they become practical problems" to be prevented or mitigated through breeding for particular traits, through spatial or bodily management, and through various other methods of 'disciplining' the subject (Gillespie 2012: 126–127; Taylor 2013: 544–545).

Likewise, very few options are available to prisoners to resist punitive practices and to reject their status as 'bare lives'. Hunger strikes, refusals to cooperate, and creating violence all are tactics that perhaps more than anything else, ensure retaliation from prison administration. That said, it is also clear that within these carceral sites there are always degrees of agency, and a continuum of power relations. Moreover, what constitutes 'violence' and 'cruelty' themselves across incarcerated individuals and settings varies considerably (Chapter 1), and prisoners themselves would no doubt view and participate in various work programs on the inside, for example, in a wide range of ways. Thus it is important to recognize that even within the vastly uneven power relations that inhere within the carceral sites of the prison and the farm, disciplinary practices and forms of resistance to them will configure in diverse ways and to varying degrees (Emel et al. 2015: 171; Rhodes 2009).

Much more could be said about agency and resistance within carceral spaces (see Chapter 5). But by way of concluding my discussion of human and non-human property and commodity relations I would like instead to turn briefly to the provocative fact that the processes of labor exploitation in carceral spaces inexorably transform and change prisoner and animal bodies. Carceral spaces not only can inflict pain and suffering on laboring bodies but they necessarily 'transform' them as well: what happens in the space can never be undone, the subject can never be transformed back. This raises the question of what types of 'redemption' are possible within the calculus of property and commodity relations. If carceral space does not kill – e.g. transforming a sentient being to a commodity such as a piece of meat or lab specimen – the action performed within carceral space will nonetheless change the subject, transforming a wild creature to domesticated or rescued animal (Collard 2014), or transforming the ostensible 'criminal' to redeemed citizen by forcing adoption of normative Whiteness or reform principles via educational or vocational training, wherein the prisoner thus comes to represent a new commodity (Shabazz 2015b).

As discussed above, there appears to be little to no evidence that prisoner work programs on the inside rehabilitate prisoners or prepare them for jobs or life on the outside. The 'achievements' of these work programs, rather, accrue as profits primarily to the state or private enterprises. Thus it behooves us to question what happens to the body, the disposition, the consciousness of the further disenfranchised prisoner-laborer? Gordon (2011: 15–16) asserts that in order for prisoners to achieve a measure of redemption within this context, that they 'master the present' and 'forge a relationship to futurity' itself. This means refusing to 'serve time' and instead devote a prison sentence to restoring civil life and citizenship. To her this means becoming members of social communities, finding meaning in art and literature, in radical thought or in organized resistance and rebellion, "which is to redeem a future, a life, out of a space of living death."

Narratives of redemption and "second chances" pervade the representational space of the prison for both humans and nonhumans alike. Glick (2013: 652–653) offers an example of a celebrity dog-fighting scheme, a reformed prisoner, and rescued dogs.[6] McKee (2015: 38, 46) posits a similar example of prisoners and ex-racehorses 'defying' their death sentence, given a second chance, and becoming "redemptive capital" together – lives worth saving, whether horse or prisoner. McKee asks, once a life is rescued from social death (or actual death), what becomes of its so-called second chance once the threat of death has been eased? Can life become more livable, supported, and enabled – redeemed – when it is rescued … or does the mark of death and the conditions of killability that lead to it still haunt any second chance at life (2015: 40)? Drawing on Gordon (2011) and Butler (2009), McKee (2015: 49) argues that redemptive capital's power lies in its ability to call forth and channel "specters or ghosts" that notify us

> that what's been suppressed or concealed is very much alive and present, messing or interfering precisely with those always incomplete forms of

containment and repression ceaselessly directed towards us ... [the prisoners and horses] enter different ecologies of care and relationships, they thrive in new and renewed ways – this is their redemptive capital, allowing them to advance from one less precarious social position to another.

Critical animal studies scholars such as Collard (2014), Blue and Alexander (2015), Tuan (1999), and Srinivasan (2013) offer useful frames for considering the 'de-commodification' of animals, the potential for animal lives to be reha-bilitated, redeemed, or 'rescued'. In Collard's example, 'wild' animals' re-entry into the wild depends on making them fear and hate humans; nonetheless they are never 'unchanged' even if returned to the wild.[7] Thus as Gillespie and Collard (2015: 9) suggest, 'multispecies justice' might be found in distance rather than proximity and intimacy, in difference more than similarity. Such is a key insight, and one whose ethical nuances and implications I take up further in the Afterword.

Notes

1 In 1987 Cabana (1996), not incidentally, as warden, oversaw two state executions by lethal gas at Mississippi State Prison.
2 How, why, and where animals work for their own benefit is not part of my discussion as these fall outside of what I consider 'the carceral'. As geographers such as Emel et al. (2015: 164–166) and Smith (2014) observe, animals perform 'work' of their own accord, building nests, acquiring food, and so on. These authors thus question various farm policies and operations that disable versus enable animal 'agency' in farm work, suggesting the possibility that humans view nonhumans as cooperators and collaborators on the farm. Yet basic to any 'revaluing, rethinking, and living another way' with animals on the farm must be the recognition of the instrumentalism and inequalities involved in human–nonhuman working relationships (Haraway 2008: 73–74).
3 Wild animal populations could, though, be considered 'free-living' property of the state, in the sense that they are managed by and controlled by state agencies, and the state issues hunting permits and licenses to trap and kill certain species in certain numbers (see Deckha 2013a: 787).
4 In 2008 the DOC issued a policy statement declaring its intention to put all prisoners to work, except those medically cleared; yet there were not enough jobs for the number incarcerated so prisoners instead received a nominal compensation until such time as work became available (the lack of sufficient placements remains the case). Moreover, because the program is designed for a successful reintegration upon release, lifers employed could not exceed 10 percent of the workforce or lifer population (DOC Policy Statement: Inmate Compensation 2008). As one incarcerated individual wrote to a local nonprofit, the Lewisburg Prison Project, "Generally speaking there are long waiting lists for many of the job assignments. The basic pay rate is .19 cents per hour (comparable to what *Haitian* laborers receive). And from there you may be able to work your way up to .42 per hour ... The DOC has not added any increase to its payrate for inmates since the late 80s. Prior to that, whenever a new Commis-sioner came into office, there was usually a *penny* or *two penny* raise of the payrates" (Anonymous Pennsylvania prisoner 2017). The Pennsylvania DOC's newsletter, *PCI Correctional Industry News*, offers an overview of current enterprises, one of which is a laundry program employing 258 prisoners who processed 17,732,797 pounds of laundry at four locations in 2015–2016. The Inmate Laundry Training

Program offers coursework that leads to "nationally recognized certifications," including "Certified Washroom Technician" and "Certified Linen Technician," certifications which 97 prisoners have completed to date. As the DOC claims, "upon release these inmates have real life laundry knowledge and experience which will provide opportunities for employment at industrial and institutional type laundries" (PCI 2017). Finally, it is worth noting that in 2017 a bill was introduced into the Pennsylvania legislature to establish the "Prison Industry Enhancement Authority" to increase joint ventures between correctional facilities and private industry. The bill declares that, "private industry in this Commonwealth will become more competitive in the marketplace while not displacing job opportunities for civilian labor in the community" (Senate Bill 2017).

5 Kim (2017: 37) illustrates the racialization of this calculation by questioning why the killing of hundreds of unarmed Black people by police in recent years does not enter the realm of 'murder'.

6 The outrage on social media over the 'man who kicked the cat' on the streets of Brooklyn and was later arrested for it (Stepansky et al. 2014; Schlossberg 2014) says much about the deep contradiction in U.S. social life not only about the 'proper' places for abuse and torture of nonhuman animals; but the fact that the man in question was Black should also not go unnoticed. As Kim (2011) and Glick (2013: 644–646) argue, such outrage parallels that over football legend Michael Vick's dog fighting scheme, outrage that had more to do with Vick's Blackness than with the animal abuse itself. The media, activists, and legal forums all referred to Vick's "execution" of dogs by electrocution, drowning, and hanging following the end of their usefulness in dog fighting. In what contexts can nonhuman animals be executed? Executability depends on particular definition of 'matterable' life, one that has political and cultural meaning. Vick's dogs, then, were elevated to quasi-human status while Vick himself became more 'animal', an outcome of the long history of the racialization and criminalization of Black men – the 'spectre of black male violence'. The experiences of both Vick and the dogs followed a narrative of redemption as he was released from prison and the dogs were rescued.

7 Blue and Alexander (2015: 158–160) offer another example of human co-existence with non-domesticated carnivores, specifically human interventions in adapting to the presence of coyotes in a Toronto community. Among the many coyote conflicts reported in Canada from 1998 to 2010, the story of the coyote 'Neville' illustrated that through practices such as securing garbage, keeping children and small pets close at hand, and monitoring the accessibility of yard plants, Neville was given a 'stay of execution'.

5 Wildspace

The cage, the supermax, and the zoo[1]

Introduction

FIRST WOMAN: It is such a pity the circus is closing! I can't believe the circus is closing!

SECOND WOMAN: It is, such a pity! I'll miss the animals, the elephants.

FIRST WOMAN: The animals, they have such good lives in the circus, enjoyable lives.

SECOND WOMAN: I would be opposed to the circus if the animals weren't taken good care of. But they're taken good care of in the circus, they're taken really good care of.

The above conversation was one I unavoidably overheard recently at a local nail salon, about the closing of the Ringling Bros. Barnam and Bailey Circus – the "greatest show on earth" – in May 2017, after 146 years of operation (Mele 2017). The conversation belies what animal rights activists have long observed about the cruel and unnecessary treatment of animals at the circus, particularly elephants, throughout the long run of this 'once beloved' form of entertainment for humans (Urbanik 2012: 80–81). At the circus, like the 'beloved' zoo, the water park, and other spaces of animal entertainment, animal bodies are commodified, treated as property, and their labor is exploited for the profit of others – and even if such spaces ostensibly exist for other purposes such as educational and scientific inquiry (below). One might observe, along with Collard (2014: 153), that in the animal entertainment industries, a particular mode of value-generating life predominates, where the value of animals is based not on their deaths, but rather on their "liveliness, wildness, and 'encounterability'."

These spaces of animal encounterability also rely fundamentally on particular forms of 'caging'. Thus in this chapter I bring forward a discussion of 'the cage' from a number of related angles, for both nonhuman and human subjects. I focus specifically on caging subjects and practices within the zoo and zoo-like structures, although certainly further inquiry into elephant and other animal caging within the circus is warranted (Yoram 2002; Tait 2011; Nance 2013). I juxtapose the zoo and the supermax prison's solitary confinement cell primarily because such a juxtaposition helps us understand the carceral logics that respectively underpin these carceral spaces, including how caging both humans and nonhumans requires producing them as 'animalistic' first. Comparisons across these carceral spaces offer a number of opportunities to study the relative histories, practices, experiences, and politics of caging. These include the historical-geographical dynamics of the cage and underlying disciplinary regimes of the zoo and prison; the cultural and sociological 'mandates' of caging; the associated psychological-behavioral experience of being caged; and the political and ethical challenges to long-term captivity in cages that the respective animal rights and prisoner rights movements have brought forward.

Much has been written about the penal philosophies, conditions, economics, and politics of caging humans in prison solitary confinement or lockdown cells (Haney 2008; King et al. 2008; Mears and Reisig 2006; Guenther 2013). Scholars – and of course prisoners themselves (e.g. Mulvey-Roberts 2007) – understand these sites as locations of civil and human rights abuses. In a different context, many animal rights scholars and activists challenge the conditions, ethics, and damage caused to animals confined in small cages in zoos and in other facilities employing similar forms of captivity (Jamieson 1985; Kemmerer 2010; Acampora 2010). While I acknowledge the risks involved in making comparisons, and appreciate that great care must be taken to do so, in my discussion I note a number of parallels that can be drawn between how human and nonhuman animals alike experience and/or act upon such spatial tactics of enclosure.

Figure 5.1 Elephant in a cage. Photograph courtesy of PublicDomainPictures.net.

Obviously there are crucial differences between the purpose and function of the zoo cage and that of the prison cell; animals held captive in zoos are not being punished for individual criminal transgressions (even if, as Girgen 2003, Nast 2015, and Glick 2013 point out, nonhuman animals do continue to be punished in other spaces as if for criminal activity; see Chapter 1). As I highlight below, the rationale behind the institution of the zoo has gradually shifted in recent years, with ostensible education and promulgation of ecological objectives replacing the increasingly objectionable purpose of simply providing human entertainment (Braverman 2013a: 86–91; Acampora 2010). But regardless of the purpose of the zoo or prison, we can observe many structural, functional, technological, and experiential similarities across these sites and institutions.

In many ways the crisis of hyper incarceration, and the human rights questions posed by increased use of solitary confinement and/or permanent lockdown in maximum-security prisons, map onto the development of the zoo and debates about caging animals for human entertainment and research. In this chapter I explore and compare the movements and rights discourses surrounding each, examining their effectiveness, relative successes, and roadblocks. What, if anything, can be learned by cross-pollinating these discursive and activist fields in attempts to advocate for both prisoners and animals? I take up these questions in the sections that follow. And I acknowledge that there are, of course, many forms of caging humans and nonhuman animals that I do not discuss here directly but have done so indirectly in previous chapters, and which have broad relevance to my topic. Animals caged in biomedical and other research labs, despite protections especially for primate animals, remain at a critical crossroads (Chapter 3); and animals captive within the cycles of industrial agriculture experience some of the most horrific conditions of the cage (Chapter 4).

Comparisons can be made between prison cages and other types of human enclosures (asylums, camps, and so on; Malamud 1998; Watts 2000; Loyd et al. 2012; Moran et al. 2013), and such comparisons are helpful in thinking through other ways that nonhuman animal and human enclosures can be studied together. However, I confine my attention to zoo- and prison/jail cages here because study of their similar geographies, disciplinary regimes, and rapid transformations over the past 40+ years provides an opportunity to examine their respective ethico-political challenges that can, in turn, speak back to theories that inform both critical animal and human geographies. I then pick up these themes further in the Afterword.

The caged body

> His vision, from the constantly passing bars,
> has grown so weary that it cannot hold
> anything else. It seems to him there are
> a thousand bars; and behind the bars, no world.
>
> As he paces in cramped circles, over and over,
> the movement of his powerful soft strides

is like a ritual dance around a center
in which a mighty will stands paralyzed.

Only at times, the curtain of the pupils
lifts, quietly. An image enters in,
rushes down through the tensed, arrested muscles,
plunges into the heart and is gone.
                    The Panther, Rainer Maria Rilke (1982)

I'm deeply cornered in their prison. My sight is diminished, but I maintain my vision ... I see forced feedings, cell extractions, mind medications, and chemical weapons used to incapacitate. I see a steady stream of petty hassles, harassments, verbal barrages, mind-fuck games, disciplinary reports, medical neglect, and the omnipresent threat of violence. Airborne bags of shit and gobs of spit become the response of the caged. The minds of some prisoners are collapsing in on them.
                    (Raymond Luc Levasseur, 'ADX the First Year' 2005: 48)

The caged bodies of zoo animals and humans confined in maximum-security prisons can be studied through a number of rhetorical, material, and physiological parallels. They both can be situated as spectacular, wild, and dangerous bodies that 'require' enclosure; as victims of the physical and psychological abuses of enclosure; as oppressed bodies wholly without social or political rights – as homo sacer – 'bare lives' (Agamben 1998; Rhodes 2009); and as representatives of the capital accumulation strategies of zoos and prisons (Harvey 1998; Nibert 2002; 2013). As Malamud (1998: 117) succinctly observes, when we make the institutional parallel of the zoo to the prison it is always the body that is at issue: "the body and its forces, their utility and their docility, their distribution and their submission." As Haney (2008: 969) adds,

The comparison [of the prison] to a zoo ... is apt – places where exotic, presumably dangerous species are caged in so completely, far from their natural environment, kept separate from one another and largely apart, even from their keepers. The haunting similarities are many in number, and one is hard-pressed to name any other place in our society where sentient beings are housed and treated this way.

Particularly important are the intersecting social constructions of the human-nonhuman caged body as wild and dangerous, with the isolation of the cage intended for the ostensible protection and safety of others. Caging humans requires producing them as dangerously animalistic first (Wacquant 2001; Rhodes 2009). The cage connotes the need to restrain a beast who is lacking in self-control, a human (usually male) who is seen to behave like a dangerous, brutal, coarse, cruel animal (Malamud 1998: 114). As discussed in previous chapters, it is the social construction of 'the animal' – the social meanings attached to various marginalized or vilified groups – that is used to perpetuate hierarchical human–human as well as human–nonhuman relationships, with racial difference basic to much of the 'criminal as animal' rhetoric.

Malamud (1998: 106) discusses the ways that early zoos appealed to a taste for horror by stressing, and exaggerating, the savagery of their animals. Braverman (2013a: 6–9) argues that the most crucial assumption underlying the entire institution of animal captivity is the classification of zoo animals as wild and therefore as representatives of their unconfined conspecifics. "Take this assumption away, and you take away the raison d'etre of the zoo." Jamieson, though, in his classic "Against Zoos" (1985), describes the profound denaturing that occurs at zoos; while animals may look like their wild counterparts and share the same genetic code, they lack the behaviors, skills, and awareness of those in natural habitats. Breeding animals for zoos or experimentation, for example, while not the same thing as complete eradication of a species, can be tantamount to annihilating it by imposing social death on its individual members (Davies 2013; Braverman 2013a. Of course there are many species 'in the wild' threatened with literal extinction too; Braverman 2015).

Acampora (2006: 104) describes zoos as thus producing a "generic animality." Zoos unquestionably portray their animals as wild and untamed, and intimately related to those in the wild and thus sharing in their plight. In attempts to maintain the wildness of zoo animals, many zoos have ceased giving them Western names. Today they are given a number (like the prisoner), or an African name, further locating them as undomesticated. Timmy and Helen have become Mshindi, Kweli, or Mia Moja, names the public will presumably associate with 'wild' and distant Africa, even if they were born in the U.S. (Braverman 2013a: 9).

Prisoners themselves turn to animal imagery to express the dehumanizing effects of isolation and exposure in the prison. Many express shame and anger at being caged in view of others like animals in a zoo. As one prisoner who Rhodes (2009: 197) interviewed in a Washington State prison described it, "if you choose to put these people in a box with nothing, what you're gonna get out of that is stark raving animals. I've seen animals produced in this very hall. People who have just lost their total cool." An Academy Award nominated documentary, *Doing Time: Life Inside the Big House*, covered the humiliations and rank conditions at a federal prison in Pennsylvania, USP-Lewisburg, including some poignant scenes of guards referring to the place as a zoo and prisoners responding by 'woofing' like dogs or wolves for the cameras, and then declaring, "we're dying" (Raymond and Raymond 1991). The cultures of many prisons reinforce this relationship. At Douglas County Correctional Center in Omaha, Nebraska, where I conducted research on prison spatial design and violence, the unit where the most at risk or unstable prisoners are isolated is colloquially referred to as 'the zoo' by staff and prisoners alike (Morin 2016).

The spectacle of the wild

A significant tension can also be noted between the visibility-invisibility of both caged animals in zoos and humans in prisons. We see a kind of violence and disempowerment in the perpetual visibility of both, and that visibility – combined with the social invisibility or secrecy surrounding the practices that control and regulate human and animal subjects – is integral to their respective animalization and objectification.

To be viewed, to be something to look at, is an integral part of the caging process (Benbow 2000; 2004; Acampora 2006: 102–115; Berger 1980). Zoo cages – public spaces at the heart of the urban center – are designed *as* spectacle. Perhaps nowhere can we observe this carceral logic better than in current practices in Danish zoos which euthanize and then dissect animals 'superfluous' to the gene pool, for the media and public consumption – illustrating not just the killability of the animals but emphasizing their utter disposability as well (Bilefsky 2015; Parker 2017; Braverman 2014). While the zoo by its very nature requires visibility, many or most animals are in fact actually hidden from view much of the time, when the zoo is closed and also when they are rotating on and off display (DeGrazia 2002; Braverman 2013a).

As Acampora (2006: 110) and Braverman (2013a: 86–92) posit, modes of carceral visibility have productive effects going beyond simple suppression, suggesting a relation of sorts between the disciplining mechanism of the prison panopticon and its parallel, the "zoopticon," with respect to the gaze, self-discipline of the subject, and the spectacle of confinement. In the logic of the zoopticon, visibility is the key concern (Acampora 2006: 110):

> the function of controlling the animals' placement and diet is to habituate them to tolerate indefinite exposure to the visive presence of humans. The zoopticon is a kind of panopticon turned inside out … to produce the same result: an institutionalized organism largely incapacitated for life on the outside.

Even if animals cannot be the fully realized subjects of disciplining with a comparable incentive for behavioral change as in the prison, where the panopticon functions to create docility through the power of deterrence and shame (Foucault 1977), clearly zoo animals are rewarded and/or punished for their performance, docility, aggression, and successful (or not) role as zoo spectacles. Meanwhile, as (Braverman 2013a: 88–89) argues, the viewing public is also being 'disciplined' in the zoopticon and indeed created; a new public inscribed with "new relations of sight and vision, educated into a certain philosophy of nature and conservation, care and sympathy."

Most prisoners, especially those isolated in maximum security, are hidden and secreted away, outside of public view, though at the same time they are subjected to constant, uninterrupted surveillance within prison walls. When their caged experiences turn into a spectacle for outside audiences – whether for television or other film/video media outlets, prison tours, museum histories, or live performances – the association of the wild and animalistic is oftentimes reinforced (Schrift 2004; Turner 2013; Morin and Moran 2015).

Montford (2016) offers a useful analysis of debates surrounding the prison tour – tours which exploit prisoner bodies as objects and spectacle for outside visitors, and yet which also provide an important access point to spaces otherwise inaccessible to the public. Montford observes that it is a very common trope

Figure 5.2 Prisoners recreate in crowded conditions at Orleans Parish Prison yard, New Orleans, Louisiana. Photograph courtesy of Bart Everson, 2012.

of prison tours that prisoners are viewed as animals in the zoo; and because both those opposed to and those in support of prison tours focus on the 'dehumanizing' indignities experienced by prisoners as if they were animals in a zoo, they equally reinforce the animality of zoo animals and thus human superiority. Again, as many such as Kim (2015) argue, this 'animalizing' process creates the hierarchical conditions that allow for the exploitation, abuse, and disposability of both certain nonhuman and certain human populations in these carceral settings in the first place. And as Guenther (2012: 56–58; 2013: 155) adds, what the

> opposition between humane and inhumane treatment fails to grasp is the degree to which it is not primarily as *human beings*, with a presumably inherent sense of dignity and freedom [that individuals are damaged] … but as *living beings*, sensible flesh, with corporeal relations … It is as *animals* we are damaged or even destroyed by [solitary confinement] just as our fellow animals are damaged or destroyed by confinement in cages at zoos, factory farms, and scientific laboratories.

Caged: response and resistance

The practice of caging elicits many empirically recognizable behavioral and psychological responses in both humans and nonhumans, including, to varying degrees and extent and in different ways, depression, despair, lethargy, stress, fear, shame, and eventually anger, acting out, and violence. These responses have the tendency to reinforce preexisting assumptions that the enclosed body is wild, bestial, and savage, thus requires caging. It is important to keep in mind though that animal experiences and responses to caging will of course vary by species, individual animal, and type of 'normal' habitat (after King 2013). That said, zoo critics nonetheless make the generalizable assertion that confining animals in cages produces anxiety, sadness, neurotic behavior, poor hygiene, and suffering (Jamieson 1985; Acampora 2006, 2010; Francione 2000; Rudy 2011; Braverman 2013a; Berger 1980: 24–28). This is true even for those under the care of the best-intentioned zookeepers, let alone those who neglect or mistreat animals. While some animals existing in such captivity are arguably well fed, disease-free, and comfortable (DeGrazia 2002: 92), their living environment is boring and physically stifling, their "jailhoused bodies" never fully able to engage their physical and mental capabilities (Acampora 2006: 99–103).

Zoo animals typically lack companionship, adequate exercise, and stimulation. As Jamieson (1985: 109) asserted, zoos can never hope to provide experiences that animals deserve – gathering their own food, living in social groups, behaving in ways that are natural to them. While Rudy (2011: 126) observes that most zoo animals today "reveal no outrageous behaviors such as self-mutilation or obsessive behaviors such as spinning or pacing," this is likely because animals who engage in those behaviors are simply taken off of display. Conversely it is the case that most zoo animals today were born in captivity and thus many zoo proponents argue that the cage is a normal, comfortable experience for them (Zimmermann et al. 2007).

Human responses to caging in maximum-security lockdown and/or isolation in prisons have been well established in the psychological and criminological literature. Mears and Reisig (2006: 34) assert that the supermax differs from earlier prisons with lockdown 'holes' in that they do not aim, even at an ideological level, to reform prisoners; rather, they are intended to 'break' them through isolation. In the supermax, prisoners eat, sleep, live, exercise, and die in their cells alone or with a cellmate. Haney (2008: 956), Mears and Reisig (2006), King et al. (2008), and Rodriguez (2005) among many others describe the oppressive day-to-day prisoner experience: it is one of overall sensory deprivation, isolation and loneliness, enforced idleness and inactivity, oppressive security and surveillance procedures, and despair (cf. Levasseur 2005, above). Those not broken by the system may become more dangerous and mean. Violence or the constant threat of it is one guaranteed by-product of the supermax, and mental illness the other. The supermax 'manufactures madness': "prisoners subjected to prolonged isolation may experience depression, rage, claustrophobia, hallucinations, problems with impulse control, and an impaired ability to think, concentrate, or remember"; approximately 90 percent of this imprisoned population suffers mental illness (Magnani and Wray 2006: 100; Guenther 2012: 50).

Guenther argues that for humans and nonhumans alike, 'everything is different' in the solitary cell or wire or concrete cage of the prison, factory farm, zoo, and laboratory; such confinement creates a particular kind of skewed relationship between the self and the surrounding world of objects. She argues that other scholars have not quite captured the feeling or experience of "living death" in a cage – that feeling of becoming "unhinged" from reality when deprived of everyday encounters with other living creatures, when subjectivity itself is broken apart and "even one's sense of individuated personhood threatens to dissolve" (Guenther 2013: xii–xvii; also see Dayan 2011). As Guenther (2013: 145) describes such experience:

> [The individuals'] senses seem to betray them; objects begin to move, melt, or shrink of their own accord. Even the effort to reflect on their experience becomes a form of pathology ... they cannot think straight, cannot remember things, cannot focus properly, and cannot even see clearly. ... even to tell where [their] own bodily existence begins and ends.

Rilke's (1982) poem *The Panther* in the epigraph above captures this skewed relationship of the captured zoo animal and the outside world, the damaged corporeal relationship experienced between the inside and the outside: there are only bars and 'behind the thousand bars, no world'. For both humans and non-humans alike, "there is a network of relations with other living and nonliving beings that helps sustain a meaningful sense of being-in-the-world," and moreover there is nothing intrinsically human about this need for everyday experience that is fundamentally relational and inter-corporeal (Guenther 2013: 149–157).

In the prison, unless receiving a diagnosis of mental illness (if one can somehow display 'normal' resilience to these conditions of confinement) – a prisoner can be subjected to the very conditions that produces mental illness. Moreover, the Eighth Amendment protections against cruel and unusual punishment do not apply to such conditions of confinement (Dayan 2011; 2005: 50). That is, the 'cruel and unusual' limitations are only applied to clinically identified mentally ill prisoners, and thus in an insidiously circular manner, mental illness is the benchmark for distinguishing torture from legitimate punishment, but is also that which is produced by that 'legitimate' punishment.

There are also a number of parallels for how both human and nonhuman actors attempt to resist caging practices and enclosure, acting on their conditions with gestures and acts (Rhodes 2009: 199). These are agents who oftentimes have no other means of resistance other than using their own bodies as weapons. As Braverman (2011: 1700–1701) argues, "despite their subjugated legal position, animals are nevertheless active subjects embodying a form of agency in their ability to continue to challenge, disturb, and provoke humans." The recent documentary *Blackfish* (2013) testifies to the violent behavior of SeaWorld's orca whale Tilikum, produced as a result of prolonged captivity in small tanks. News stories and television programs such as the popular *Animal Planet* frequently play up the horrors and 'irrationality' of zoo animals attacking their human caregivers as such (e.g. Gates 2013; see Gillespie 2012: 122–127 for additional examples). Yet when animals kick back at or attack their human caregivers perhaps they are exercising their own 'natural laws' (also see Philo and

Wilbert 2000: 14–23). In such cases it is useful to keep in mind that such resistance is necessarily coded and filtered through human language and understanding. Kim (2016: 41–42; after Derrida 2002; Haraway 2008: 77–82) reminds us that 'independence, resistance to human sovereignty' can be inaccessible to us as either reaction or response; and in fact, "we have good reason not to try" to decipher the animal mind in such cases since wild animals – in her discussion, Harambe the gorilla – are not "reliably submissive" and obedient (see Postscript below).

Much could also be said about the few opportunities that isolated prisoners have to resist their spatial enclosure and associated punitive practices. Hunger strikes, refusing to cooperate, using their bodies as weapons, and creating violence are all tactics that carry numerous risks and ensure retaliation from the prison administration. Throwing food and feces, acts of self-mutilation, biting, destroying material surroundings, and so on are unsurprising outcomes of desperate individuals ensnared in the perverse and violent supermax system of punishment (James 2005). Such tactics and agency, while challenging the mute passivity of the imprisoned also, though, further reinforces preexisting associations of prisoner animalism.

This brief overview of some of the resonances across the caged bodies and experiences of zoo animals and prisoners offers a useful context for examining the historical transformation and carceral logics of the zoo and prison as carceral spaces of enclosure. I turn to those now, and subsequently to the various rights movements that have influenced the development of these spaces, especially over the past 40 years.

The zoo cage

Zoos are among the most popular cultural institutions worldwide; approximately 100,000 of them attract over 600 million visitors each year (Zimmermann et al. 2007: 4; Bulbeck 2010: 85). Today there are many kinds of zoo, and they vary considerably in quality and purpose, ranging from small, bleak 'concrete prisons' to naturalistic, conservation-oriented bioparks that attempt to replicate animals' natural habitats (DeGrazia 2002: 88; Rudy 2011: 122; Acampora 2010; Berger 1980).

The zoo as a site for modern entertainment, education, or scientific study emerged in continental Europe in the 18th century, and in the U.K. and U.S. in the 19th century. Hallman and Benbow (2006) explain the evolution of Western zoos in three distinct stages (also see Acampora, 2005; 2006; 2010; Uddin 2015). The early 'menageries' were dedicated to entertaining the public, with rows of bare cages enclosing single specimens and intended to reinforce "notions of human power and superiority over the natural world" in the age of colonialism and empire (Hallman and Benbow 2006: 257). Fast-forward to the post-war era, and the 'living museum' began to emerge, with enclosures built to resemble jungles and woodlands. Such spaces were meant to "banish the emotional response to human dominance over less powerful animals," emphasizing ecological relationships, habitat and species conservation, and public education. The late-20th century "conservation centre" appeared as the third stage, a place that "exhibits active concern about the exploitative relations humans have with animals" and thus brings human visitors "inside the cage." Protecting biological diversity and sustainability are central to these later institutions (Hallman and Benbow 2006: 259–261); including, as Uddin (2015) argues, as part of an attempt to counteract urban decay itself. Samuels (2012: 33) adds a fourth stage of zoo development to this typology, a mode of display he characterizes as "eco-tainment," in which "giddy amusement-park tricks offer a measure of relief from the knowledge that nature is only another man-made illusion." Beardsworth and Bryman (2001: 91–98) call this the "Disneyization" of zoos, involving a combination of "theming," consumption/merchandizing practices, and the emotional labor of zoo workers.

Such developmental stages can and often do co-exist at any given site, as my recent visit to New York's Bronx Zoo confirmed. Widely acknowledged to be one of the world's 'best' zoos, the Bronx Zoo of today dedicates itself to "saving wildlife and wild nature," and many of its exhibits are framed with information about conservation and endangered species (such as the reproduction of a poacher's truck on Tiger Mountain). Nonetheless the Dinosaur Safari, Bug Carousel, and the Wild Asia Monorail, all aimed at having fun, obviously distract the zoo visitor, perhaps particularly children, from the more painful reminders of animal abuse and habitat demise (Bronx Zoo: Experiences 2017).

It is clear that in the last 40+ years we have witnessed the emergence of an era of the benevolent or ostensibly 'progressive' zoo – an obvious reform of zoo conditions and a change of mission. Today, many zoo advocates argue that at least for larger primates and mammals, small concrete barren cages have generally given way to larger, more naturalistic habitat enclosures, some components of which are made of natural materials; animals are rarely displayed alone; and zoos attempt to educate the public about endangered species and habitat destruction noted in informational signage. Zoos vary tremendously on these features, however, and regulatory oversight as well as collection of empirical evidence to support such assertions has been unsystematic to date (Zimmermann et al. 2007; Braverman 2011; 2013a).

Debates about zoo cages today center on whether these progressive sites are in fact "a new and acceptable form of wild animal keeping, or whether they are simply a dressed-up version of the colonizing, concrete prison model" (Rudy 2011: 123–124; Acampora 2006: 103–115; 2010: 1–8). As Benbow (2000: 13–15) argues, technology and culture have conjoined in the development of new mechanisms for caging captive animals – including various forms of wire, glass windows, electronic fences, walls, ditches, and moats. She explains that the most significant change that has occurred in the geography of the cage is the (larger) size of enclosures. Nonetheless, in zoo architecture, a balance or compromise is always made between the conflicting "aesthetic demands" of visitors and the needs of the animals. Thus, rocks, vegetation, and other spatial features, for example, whether synthetic replica or the real thing, are provided primarily for the visitor (not the captive animal), intended to evoke a natural environment, habitat, or themed region of the earth (Benbow 2000: 18–20; Francione 2000: 24). Thus many would argue that the impression of 'cagelessness' in this 'natural' zoo habitat is merely pretense, with animals simply subjected to more sophisticated regulation (Malamud 1998: 107; Acampora 2006: 103–108). Ultimately, what all zoos have in common is the display of animals to the human public; and what all zoo animals have in common is the experience of being observed, as object, within a hierarchical relationship with the observer.

Moreover, zoos arguably do nothing to address the primary causes of global biodiversity loss, unless one considers the intangible benefit of education, which itself is highly debatable (e.g. Braverman 2013a: 18). Zoo critics such as Francione (2000: 25) argue that watching a lion in a zoo is no more beneficial than watching a film of a lion. Others, however, argue that witnessing the embodied "wonder, beauty and mystery" of zoo animals has great potential for changing public attitudes (Benbow 2000: 15; 2004: 379; Zimmerman et al. 2007: 4–7). As Jamieson (1985: 111–112) noted though, "undoubtedly some kind of education happens in zoos, but the question is, what kind? … couldn't most of the important educational objectives better be achieved by exhibiting empty cages with explanations of why they are empty?"

Many also argue that zoos have not been successful at maintaining genetic diversity of endangered species and/or re-introducing species back into the wild, due to the substantial financial, health, and adjustment risks involved. Acampora (2006: 106) questions the utopian vision of "Releasement Day," arguing that this sort of futuristic planning presumes unrealistic, wide-scale, socio-ecological stability over a very long time; meanwhile Francione (2000: 24–25) describes the inefficiency, waste, and abuse inherent in breeding programs. Most zookeepers readily admit the obvious, that the best way to 'save' wild animals is to protect their habitats and sanctuaries (Rudy 2011: 119; Kemmerer 2010: 42). But because maintaining actual natural habitats and protected areas has not proven feasible, especially for large land vertebrates, the zoo becomes an assumed 'necessity' to breeding endangered species and protecting biodiversity. Zoo advocates argue that modern zoos emphasize conservation, education, and care and stewardship of animals as their central mission; that in caring for animals in zoos and breeding offspring, they are caring for animals in the wild, i.e. "saving" wildlife (Braverman 2013a: 15–17; Zimmermann et al. 2007). As yet again Acampora (2006: 112) argues,

> Despite the best intentions, efforts, and achievements toward making these forms of confinement more comfortable for the inhabitants and more palatable for visitors, zoos still share deeply in an order of prison-like institutions partially constitutive of urban modernity. Violent seizure, forced captivity, thorough exhibition, programmed feeding and breeding, commodified exchange – all these activities testify: to display animals is already to discipline them; to preserve species it is necessary to punish individual specimens (as representative inmates).

A number of scholars and advocates argue for new models of wildlife enclosure altogether – the sanctuary, reserve, Earth Trust, zoological garden – signaling, if you will, a fifth stage in zoo development. Such models signal the abolition, rather than the reform, of what is ordinarily considered a zoo.[2] Rudy (2011: 113–114), for example, supports the privately run sanctuary, founded not on putting animals on display but on developing human relationships with them. Similarly, Kemmerer (2010: 37–42) advocates for the "nooz," safe havens for individuals misused by zoos, circuses, or science, institutions framed within the logic of reparation for previous exploitation. "Nooz will not purposefully seek out prisoners from the wild, or breed prisoners to entertain human beings," Kemmerer writes.

One such place is the largest exotic wildlife rescue facility in Pennsylvania, the family owned and operated "T&D's Animals of the World" near my home. Most of the 300 animals at T&D's – including lions, tigers, cougars, leopards, wolves, and bears – were formerly abused, neglected, or illegally owned pets, or have been discarded from zoos and other operations. Most of the animals live singly or in pairs in a half to two acres of woods that also feature enclosed shelters. Although T&D's allows weekend visitors during the summer, it is not a place designed for humans but rather for animals, the latter of whom may or may not 'display' themselves to visitors on any given day. T&D's is a good example of a facility that "contests exhibition" (Chrulew 2010: 205–206).[3]

All of this said, and despite many disagreements, it seems undeniable that a progressive social and spatial evolution has taken place over the last several decades in the politics, ethics, and care of animals in captivity – however much

Figure 5.3 Bengal tigers resting at T&D's Cats of the World wild animal refuge, Penn's Creek, Pennsylvania. Photograph by the author.

remains to be done. Real reforms are evident in the caged existence of zoo animals today, yet we also see a demonstrable further evolution, beyond reform of the zoo towards its abolition. What parallels can be drawn with the evolving practices of caging humans in prisons? Where do the narratives of these institutions converge, and where do they diverge?

The supermax prison cage

What we might think of as the social history of caging humans in long-term isolation can be traced in the U.S. to the infamous experiment at Eastern State Penitentiary in Philadelphia. Built in 1829, Eastern State is today considered the United States' most historic prison, primarily for the role it played in developing penal philosophy. Philadelphia Quakers are attributed with creating the idea for this first penitentiary, a prison designed to inspire true regret, or penitence, in criminals' hearts through complete isolation, silence, and individualized labor in cells. Eastern State was a source of debate from the beginning – Charles Dickens was one of its earliest detractors, in 1842 – yet its ideals were not abandoned until 1913 when they collided with the reality of overcrowding. The prison did not close until 1971 though, and the site became a popular tourist attraction in the 1980s (Bruggeman 2012).

From the late 19th century social reformers had sought to ameliorate the deplorable conditions in American prisons – they were overcrowded, poorly ventilated, dark, unhygienic spaces where prisoners were kept in solitary cages regardless of their crime. Most were incarcerated for non-violent crimes such as horse-theft and counterfeiting; and later, under Prohibition (constitutionally criminalized from 1919–1933), for producing, transporting, or selling alcohol, which dramatically contributed to the massive prison-building spree in the early 20th century. When the Federal Bureau of Prisons (BOP) was created in 1930, its mission was to reform this system, to rehabilitate prisoners through education, vocational training, and recreation, an approach in line with contemporary 'scientific' penology.

This ideology of reform and rehabilitation suffered a short life span, however; by the 1970s these principles had completely lost traction within the prison bureaucracy and in the courts, and stood in stark contrast to the norms and practices of everyday life inside penitentiary walls (Richards 2008). Guard brutality, overcrowding, unsafe working conditions, infrastructural deterioration, and prisoner civil rights challenges led to a breakdown in the ability of the BOP to control its facilities, and uprisings frequently occurred at many federal and state facilities. Prisons became increasingly violent places, dozens of guards around the country were killed, and numerous lawsuits followed (King et al. 2008: 146; Richards 2008: 9–10; Morin 2013: 384; Thompson 2016). By 1975 the Bureau had abandoned its concept of rehabilitation, and by 1984 the U.S. Congress passed the Sentencing Reform Act, which abolished parole for federal prisoners, guaranteeing that they must serve at least 85 percent of their prison sentences. The War on Drugs conceived in the early 1970s – with specific heightened attention to criminalizing activities in Black neighborhoods and conferring longer prison sentences to them (Chapter 3) – more than any other single cause, contributed to the hyper-incarceration trends we see continuing today (Alexander 2012). The Bureau's response to the further problems that these legal 'remedies' predictably caused, such as intense overcrowding and increased violence within prisons, led to the tortuous caging practices we see today – use of solitary confinement and permanent lockdown as a primary method of prison control. Pure punishment and retribution became the norm. Today maximum-security isolation or lockdown has become a routine and entrenched fixture in American prisons of all sizes, replacing older forms of temporary prisoner segregation.

In the U.S. there are more than 30 high-security (federal and state) super-maximum prisons, confining approximately 200,000 prisoners, and the number is growing (Richards 2008: 17–18; Morin 2013; Moran 2015a; Pratt et al. 2005). 'Law and order' arguments tend to frame the need for the solitary cage in public discourse, as advocates often claim that such confinement of the 'worst of the worst' is necessary to stem violence and keep our streets safe. In reality, estimates are that only about 5–9 percent of those in isolation cages are locked down for a crime committed on the outside (Vanyur 1995); nearly all of them have been labeled as gang leaders and/or have been accused of committing crimes (such as assault) while incarcerated.

As one USP-Lewisburg prisoner wrote about the situation there (Morin 2013: 393), "the prison provides only 55½ square feet for two grown men to live and be locked down in for 18 to 24 months straight." While the Lewisburg SMU's double-celling lockdown practices bring about their own set of stresses and violent outcomes of which few seem aware, the notorious federal ADX prison in Florence, Colorado serves as a more familiar example of just how far the caging of humans can and has devolved. At ADX, prisoners are alone in their cells at all times, they recreate alone, and at no time come into contact with another human being, sometimes for years at a time. Cells are self-contained, a spatial design that maximizes security by ensuring that prisoners rarely leave their cells. Cells are made of concrete walls, floors, and ceilings, and all cell furniture is made of reinforced concrete. Each cell has only a concrete slab bed, with a built-in storage shelf, concrete desk and seat. Each cell has its own toilet, sink, and shower. Each cell also has its own vestibule and two doors – an inner open grill that allows direct observation of the prisoner upon entering the vestibule and an outer solid door that "prevents the inmate from throwing things or firing projectiles at staff" (Vanyur 1995: 92). Most services that prisoners receive are delivered to them electronically or through the small hole in the cell door.

Only recently are we beginning to see, from the mainstream media and from the corrections industry, challenges to the abuses inherent in solitary confinement, as well as more pragmatic arguments about the efficacy and economics of the practice. Raemisch (2014), for instance, Executive Director of Colorado State Corrections, checked himself into a solitary confinement cell in order to better understand the abusive nature of the practice and the psychological damage it caused. He lasted only 20 hours, becoming "twitchy and paranoid," spending his time counting the small holes carved in the walls (cf. Guenther 2013). Raemisch concluded that confining men to small solitary cages does not solve problems, "only delay[s] or more likely exacerbate[s]" them. Goode (2013), likewise, reported that the Mississippi Commissioner of Corrections "used to believe that difficult inmates should be locked down as tightly as possible, for as long as possible." But after a rash of violence at the state's super-maximum security prison in 2007, rather than tightening restrictions he loosened them, allowing prisoners out of their cells each day, offering basketball, a group dining area, and new rehabilitation programs. In response, "inmates became better behaved. Violence went down" (Goode 2013). Ultimately an entire unit was closed, saving the state more than $5 million.

The devolution to permanent isolation or lockdown in U.S. prisons as normal, everyday practice can be attributed to a number of social, legal, and political trends noted above. But 'cracks' in this system are beginning to show that might help ameliorate or reverse the trends. At this juncture we might question the relative impact that the prisoner rights movement has effected within this scenario, and compare it to that of the animal rights movement that has similarly challenged caging practices at the zoo.

Prisoner rights and animal rights: resonances and dissonances

One of the important things to notice when comparing animal rights and prisoner rights in the U.S. is the crucial changes that have overlapped in concerning ways over the past 40+ years, reflecting the period's social and spatial 'carceral turn' (Moran 2015a). Prior to that, the humanitarian basis of Progressive Era reforms led to improvements in both prisoner and animal welfare (Finsen and Finsen 1994: 27). New organizations pushed for improved treatment of animals; and new agencies governing the caging of humans (including the BOP itself) were created, based at least in theory on reform-minded principles. At the same time, while conditions arguably improved for animals caged in zoos, starting in the 1970s conditions dramatically deteriorated for humans caged in prisons. While the zoo and the prison are obviously different kinds of institutions, run under vastly different regimes of power and intended outcomes, both have been historically subjected to pressures from outside activist organizations that have, to a greater or lesser degree, effected their reform. Yet advocates for change in both the prison and the zoo have drawn on similar ethical and bio-political arguments about the cage as a disciplinary geographical space, and as such, offer a nexus of interests that can be put to productive use for both critical human and animal geographies.

While it is difficult to isolate a unified movement advocating for the rights of prisoners as an oppressed group, it is nonetheless possible to locate the origins of the idea of prisoner oppression and rights within the Civil Rights Movement generally, particularly considering that African-American and other minoritized men and women have historically comprised a disproportionately high percentage of the U.S. prison population (currently 70 percent; Alexander 2012). To Gottschalk (2006: 165), race was the "crucible" for the contemporary prisoner rights movement in the U.S.; the race question gave birth to the most powerful and significant prisoner rights movement in the world, yet the U.S. was simultaneously a forerunner in the construction of the carceral state. These two factors are linked. Gottschalk offers one of the most comprehensive explanations for why the U.S. carceral state did not encounter a more unified opposition, a question closely tied to explanations for why prison activism has a complicated, un-unified history with few successes.

To Gottschalk (2006: 165–196), the term "prisoner rights movement" refers to a broad range of moral, political-economic, judicial, legislative, and cultural activities and institutions. Key to her thesis is that strong, well organized activism inside prisons, particularly prisoners' alignment with the Nation of Islam, the Black Panthers, and other New Left organizations, exposed the deep racism in U.S. culture with which few outsiders would align themselves, particularly as their activism included strikes, uprisings, and calls for revolution. (Even the NAACP [National Organization for the Advancement of Colored People] chose to focus more on affirmative action in schools and the workplace during this period, rather than on prison reform.) Some of these activists became household names, such as Malcolm X, George Jackson, and Angela Davis. In turn this activism created a strident 'law and order' backlash from conservative hard-liners, fed as well by the successes of victims' rights movements that perhaps unwittingly facilitated the more punitive turn in corrections and caging we see today.

Within this context the prisoner rights movement – if we can call it that – evolved as diffuse and frayed efforts, with agendas, tactics, and philosophies varying greatly from place to place and across many scales of activist organizing. Today, groups at various civic scales focus on a broad range of issues, from working to improve the conditions of confinement within prisons to abolishing prisons altogether. Agendas range from calling for sentencing reform and an end to mandatory sentencing; providing resources for legal representation; opposition to the death penalty; assistance to families of the incarcerated; self-help, vocational training, and education; religious rights of prisoners; and recidivism and reentry. A number of national-scale organizations address prisoner issues, such as the American Civil Liberties Union (ACLU), the Center for Restorative Justice, Critical Resistance, and the Sentencing Project; as do regional and state organizations such as Decarcerate PA and California's Mothers Reclaiming Our Children; as well as those more local and/or devoted to the rights of individual prisoners such as Mumia Abu-Jamal's 'Live from Death Row'. This short list does not begin to capture the vast number of organizations and coalitions active today – from grassroots and community-based groups to the more mainstream and institutionalized; from those based on prisoner organizing on the inside to those at work on the outside; and from scholarly and academic organizations to those based within the corrections industry itself.

Within this landscape are many organizations that have aspired to improve and reform conditions on the inside. To the extent that prisoner rights groups have attempted to effect tangible improvements of caging policies and practices over the past several decades – specifically those to do with the degrading practices of solitary confinement and permanent lockdown – they have done so mostly by bringing court action based on constitutional or civil rights of prisoners, typically based on violations of the Eighth Amendment to the U.S. Constitution that guarantees freedom from cruel and unusual punishment (but see Dayan 2007; above). These have had a negligible success rate in the courts (ACLU 2017). Activists have to some extent begun to influence those within the industry though, such as the National Institute of Corrections, prison architects and planners, as well as various prison guard unions. Such entities recognize, among other things, that caging practices put staff at considerable risk, as well as create further instability among persons returning to life on the outside (Morton 2008; Raemisch, 2014).

Turning to the zoo, the more recent transformations of its purpose and culture is a subset of the many changes in animal treatment, care, and use brought about most recently within the 'modern' animal rights movement, begun in the 1970s U.S. alongside other mid-century social movements. The idea that animals have intrinsic rights and/or selfhood resonated with these other movements (Jamieson 1985; Nibert 2002; 2013). The 1970s were also a decade within which animal suffering was witnessed on a new and grand scale (Finsen and Finsen 1994: 5–54). Animals became a much bigger part of medical and scientific research, and animal use in industrial agribusiness rose dramatically in the post-war period (see previous chapters). With these as impetus, a virtual explosion of animal rights organizations arose in the 1980s U.S., with hundreds of organizations springing up at the local, regional, and national scales.

Perhaps unlike in other institutions within which animals are held captive and used as human resource (the farm, the research lab), transformations within zoos owed primarily to the ostensible 'voluntary' improvements made by the zoo industry in response to outside pressures from animal rights activists (Donahue and Trump 2006: 6; Braverman 2013a). Zoos existed within the larger paradigm shift to animal rights and animal welfare, and in order to maintain (or establish) credibility, zoo keeping itself became professionally managed, with evolving cultural values, standards, and missions.

During the Progressive Era zoos were basically unregulated; zoo jobs were patronage jobs, and any improvements stemmed from pressure from local conservation or animal welfare groups. Beginning in the 1940s and 1950s, college-educated, conservation-minded biologists and zoologists began to take on roles as zoo managers (Donahue and Trump 2006: 8), and the professional organization, the American Association of Zoological Parks and Aquariums (AAZPA; later AZA, Association of Zoos and Aquariums), was codified in 1972 as a separate organization. Subsequent changes to the organization co-emerged with the national animal welfare and rights groups which directly challenged zoo practices: the Animal Welfare Institute and the Society for Animal Protective Legislation among them.

As Donahue and Trump describe it (2006: 9), "despite their commitment to protecting wild animals, these professionals became formidable opponents" of animal rights activists and fought many political battles to continue operating zoos (also see Chrulew 2010: 195). Owing to outside pressures, the AAZPA was forced to clarify and (re)define its mission, eventually adopting those of conservation, education, and breeding vulnerable species. But lacking a singular purpose – and indeed attempting to sustain their conflicting and competing recreational, commercial, educational, and species preservation purposes – the industry also hired professional lobbyists to help align zoos with various conservation and animal protection groups. Faced moreover with congressional support for the federal regulation of zoos, the newly established professional network worked hard to ensure that "new laws provide zoos with a limited right to take protected animals from the wild, while at the same time acknowledging the independent regulatory capacity of accredited zoos" (Donahue and Trump 2006: 10–11, 37). In essence, zoos became and remain self-regulating institutions, subject to industry standards they themselves establish.

As Braverman (2011: 1703) explains, though, zoos are almost "extralegal creatures" regulated through myriad variances and exceptions. Owing to the powerful status of the AZA and "almost-monopoly" over relevant knowledge, zoo animals remain largely outside the provisions of official law. As a result of "the physical and cultural nature of zoos on the one hand, and the long history of political battles between the zoo industry and animal rights groups on the other," a complex and eclectic mix of international, federal, state, and local agencies and codes regulate various aspects of zoo operations, but none address the zoo as a whole or the particular needs of most animals (Braverman 2011: 1694). Some only contain standards and care of warm-blooded animals; or of endangered species; or of particular kinds of animal breeding practices; or of only non-federally licensed operations; or have oversight only of building codes or of animals as property; and so on (Braverman 2011: 1689–1702). Indeed, most regulations governing zoo buildings are primarily aimed at protecting human accessibility and safety. Thus the regulation regime of zoos, to the extent that one exists and influences practices, owes to a complex (though limited) bundle of self-regulating codes as well as various legal statutes at various scales. These in turn owe largely to much more widespread cultural pressures, including the more philosophical and ethical arguments that have led to reform of zoo caging practices as well as the abolitionist movement. In Chapter 6 I return to some of the ethical questions posed here; for the remainder of this one, I open a discussion of these 'reform versus abolitionist' tensions as they pertain to the rights of human and nonhuman subjects caged in the zoo and isolation prison cell.

The critical nexus of human and animal geographies

Evidence suggests that the relative oppression and disenfranchisement of caged animals and prisoners are closely linked, based on a range of policies, practices, and experiences. As Watts (2000: 292) categorically asserts, "[t]he zoo is a prison – a space of confinement and a site of enforced marginalization like the penitentiary or the concentration camp." Many parallels and tensions can also be noted between the 'reformers' versus the 'abolitionists' with respect to both institutions: those who advocate for improving zoo conditions to those who wish to abolish them; and those who advocate for improved conditions of confinement within prisons to those who go beyond a discussion of civil or constitutional 'rights' of prisoners to argue for the wholesale abolition of the Prison Industrial Complex (Gilmore 2007).

At first glance it seems important to recognize the relative successes of animal rights activists in transforming the zoo compared to the relative ineffectiveness of prisoner rights activists to effect change in methods of incarceration. The zoo industry radically changed its mission, ideologies, and day-to-day practices owing fundamentally to animal rights activists whose strategies of evidence collection, public education, and savvy use of the media brought animal abuse issues to the forefront (Chrulew 2010). One of their key strategies was in publicizing to mainstream public audiences the horrors occurring in the secreted spaces of the animal cage. Lobbying efforts and notable legislative and judicial changes followed, and when they did not, organizations relied on protest and direct pressure to draw attention to animal abuses, including in zoos (Donahue and Trump 2006).

The potential impact of public intellectuals and journalists who have recently taken the same approach with respect to prison abuses cannot be overstated. Within the context of intensified use of control units, solitary confinement, and lockdown, and all their associated "sort" response teams, heavy shackles, and myriad and severe daily restrictions that have become the norm, prisoners themselves have few options to mobilize, and those who do have carry enormous risks (Morin 2013). Thus some of the tactics and strategies successful with animal rights seem essential to shifting the grounds of the debate – that is, for advocates to bring issues directly to the public since legislative bodies and the courts have been unresponsive. Activists' attention to the questionable ethics of solitary confinement is beginning to make notable difference (cf. Gottschalk 2006), effecting change from within corrections that is not unlike the process by which the AZA fought but then eventually acceded to animal activists' pressures (e.g. Raemisch 2014). (See NPR 2016 and Note 1 below for one important and recent intervention.)

Communication with and from prisoners is essential in this process. One of the greatest misconceptions about prisons today is that the 'animalistic' behavior that occurs within their walls solely originates in the individual; that prisoners are locked up because they are bad, deviant, or unfortunate people, driven to crime and trouble from some indelible social or psychological cause, and that their criminal nature will follow them wherever they go. Legislators, judges, educators, and journalists, as well as much of the public, are just beginning to understand the extent to which prisoner behavior and prison violence is a product of the carceral system, not an explanation for its need.

In many ways, though, the jurisdictional and regulatory regimes across types and scales of prisons are similar to what Braverman (2011) argues about the 'extralegal' nature of zoos and their self-regulation. While prisons reside within the U.S. legal system and thus their practices are an instrument of it, super-maximum prison conditions also manifest features of a lawless, "camp-like" space, exempt from outside scrutiny with prisoner treatment typically beyond the scope of the law (Rhodes 2009; Agamben 1998). Prison administrators have been all-too-successful at making their own rules and keeping secret what happens on the inside. This, combined with the powerful influence of prison bureaucracies at every level, and community leaders with vested economic interests in keeping the PIC 'machine' going, profoundly complicates the ability to make significant advances in prisoner treatment.

Acampora (2006: 109) meanwhile, citing the works of philosopher Heini Hediger, observes that there is a deep similarity between the professionalizing practice of zoo management and paternalistically progressive calls for prisons reform. In both cases, "one wants the inmates to feel as comfortable, as snug, and as much at home as possible." Such comfort perhaps belies nothing more than increasingly sophisticated means of regulating and disciplining captive bodies. While it is hard to deny that improvements in zoo conditions have occurred over the past few decades, the "carceral comparison" is not misplaced since both the zoo cage and the prison cage create the same result: an institutionalized organism largely incapacitated for life on the outside. Zoo critics argue that even the most sophisticated zoological garden, despite not carrying a punitive or penal intent (Bostock 1993: 63), is an institution of enforced occupancy intended to display animals. Like the prison, the structure of the zoo cage as an institution of enforced occupancy, violence, and control "ensures the production of docile bodies (or dead ones)" (Acampora, 2006: 108–109). Again, it is visibility that is the key concern: the function of controlling the animals' placement in the cage, controlling their diet, programming their breeding, and so on, "is to habituate them to tolerate indefinite exposure to the visive presence of humans" (Acampora 2006: 110). Moreover, if released, they are subjected to a "carceral milieux" of electronic tracking devices, embedded microchips, and the like used to track and control their movements; which are themselves not unlike the parole boards and other methods such as house arrest, ID bracelets, and medications that are used to control and track released prisoners.

Prison abolitionists such as Davis (2003), James (2005), and Gilmore (2007) argue that reformers' arguments are misguided in that they reinforce the (illegitimate) power of the state to cage humans in the first place. James (2005) has collected many poignant prisoner stories framed around ethical arguments and the intrinsic human dignity of prisoners. Davis (2003) argues that the prison system is so judicially corrupt, racist, socially and economically debilitating to communities, and ineffective at reducing crime and keeping communities safe, that it is obsolete and should be abolished. And rather than focusing efforts on improving prisons, she argues that efforts should be redirected to crime prevention not punishment, education and support of vulnerable populations, and community restitution for infractions, among others. That said, some in the zoo abolition movement would see key differences between the "carceral milieux" of zoo animals versus human prisoners. Some would argue that because there are cognitive differences between human and nonhuman animals, humans are capable of responsibility and hence can be held culpable for their actions in ways that even the most sophisticated nonhuman animals should not be (e.g. Steiner 2008; 2016). So while we may observe equal vulnerability within zoo cages and prison cells, as I discussed in Chapter 3, "susceptibility to incarceration" itself must also undergo a radical shift in public discourse.

The incarceration tactics of long-term isolation and permanent lockdown that have become the norm in the United States today are wholly unjustifiable not least in that they have structurally targeted minoritized men in particular (Alexander 2012). For caged zoo animals, spectator interests and enforced display can and must be replaced by what Marilyn Frye refers to as "loving sight" (Acampora 2006: 113–114). These loving eyes require the disintegration of the interests of the spectator, so we must change 'how we look' and open possibilities for not looking at all, such as that promoted by rescue spaces such as T&D's and their purposeful avoidance of a certain type of spectatorship of visitors to their facility (above).

> In the twilight of the zoo, it will be up to biologists, animal advocates, and concerned citizens alike to look ahead with new eyes, devise novel and better modes of cross-species encounter, and see them through to implementation and inevitable revision (Acampora 2006: 115).

Ultimately, comparing the relative histories, practices, and bio-politics of the cage as a disciplinary geographical space offers us crucial insights to understand and then move stridently forward in challenging the abusive conditions that span species boundaries at these carceral sites.

Postscript

During the writing of this chapter Harambe the Gorilla, an endangered lowland gorilla held at the Cincinnati Zoo, was shot and killed when a four-year-old child crawled through the barrier and fell into his enclosure. Most of the public and news media seemed to support the action of zoo officials who shot and killed the gorilla to save the life of the child, even though the child's life appeared to be in no imminent danger (Kim 2017). The previous year, by contrast, public shame and outrage surrounded the killing of Cecil the Lion in

Zimbabwe by the Minneapolis dentist Walter Palmer, a man with a history of illegal poaching and trophy hunting in Africa (Loki 2016). The dentist lured the lion out of a protected sanctuary and then shot Cecil with a crossbow, later beheading him for a trophy.[4]

As Gruen (2016b) argues with respect to the Harambe case, such are deeply traumatizing events – for the child, for the surviving gorillas, for the witnesses, for the animal care staff at the zoo, and for those of us sensitive to the plight of captive animals. In the wake of such a tragedy, it seems someone must be blamed – and the child's careless mother became one such target. But as Gruen argues, "the real culprits are zoos." In the case of the maligned dentist Walter Palmer, the reversal of blame owes to the questionable 'mastery' over majestic wild animals that the dentist displayed. So we must ask, do such examples point us towards a love of animals, or rather to a mis-recognition that we supposedly 'love' these animals despite participating in all kinds of horrific acts against them on a daily basis? The institution of the zoo itself is such a place of mastery, and yet where do we find public outcry about it?

Oftentimes there is public protest over the treatment of a single individual zoo or preserve animal – a festishization, love, and attachment to an individual – but that protest rarely extends to a whole population of the species; the rest are disavowed. Focus on such individual animals has the tendency to distract us from the horrors being inflicted upon whole populations of other nonhumans (Kim 2016). Moreover, both Gruen (2015b) and Kim (2017: 37–43) note the debates that ensued when more attention was paid to such tragedies than any number of individual Black men killed in recent years by the police. The tension in these cases then not only relates to the making of the human and the animal, but also to 'animalness' and 'Blackness' ('human-Black-animal' as Kim [2017: 42] refers to the relationship; see Chapter 3). But rather than pointing fingers at each other about inadequate or disproportionate grief at the deaths of some and not others, we might instead work to develop an 'ethics of avowal' (after Kim 2015: 18–21). This would entail an open and active acknowledgement that these struggles are linked: that we should empathize with the pain and indignities of others who are disempowered and avow, rather than belittle, their search for justice (Gruen 2015b). Part of this work clearly lies in paying greater attention to how our focus can be distracted by a single exception – recall also the public outrage against the police for killing a cow who had escaped from an Omaha meatpacking plant (Pachirat 2011: 5–15; Chapter 2) – a distraction that allows us a distraction from such deep contradictions of everyday life.

Notes

1 An earlier version of this chapter appeared in Gillespie and Collard's *Critical Animal Geographies* (Morin 2015). In 2016 National Public Radio, in coordination with The Marshall Project, undertook an in-depth investigation of the penitentiary, and after their story was broadcast a collective of 40 prisoner rights groups and other organizations, including the Southern Poverty Law Center, the ACLU, and the American Council of Churches, signed a letter to Attorney General Loretta Lynch demanding

attention to the abuses taking place there (NPR 2016). Subsequently the prison began downsizing. As of this writing, 600 prisoners remain in the facility.

2 Other post-zoo models, beyond the scope of this chapter, call for an end to animal captivity altogether, in, for instance, the zoopolis (Wolch 1998).

3 Nonetheless and quite obviously, the animals at T&D's remain enclosed and captive. Moreover, the facility is essentially self-regulating: while it is licensed by the U.S. Department of Agriculture and U.S. Game commission, it is not accredited by the AZA – which, to its owners, would only mean a great deal of retrofitting for human visitors. While this particular sanctuary appears on the path towards zoo abolition, perhaps any number of animal abuses could occur at similar such facilities.

4 How the media portrayed the 'human' and the 'animal' in this example is worth repeating here. The dentist who killed Cecil – along with the local hunters and poachers – were surely cast as the 'animals' to the 'humanized' elephants and rhinos, yet these attributions are always fluid and subject to reversals. When elephants or rhinos (or other nonhumans) kill humans their 'animal savagery' is emphasized so that retribution will be acceptable.

6 Afterword

Reflections on trans-species rights and ethics

Introduction

The aim of this book has been to develop a trans-species carceral geography that offers insights into how and why the sites and spaces of human carcerality, and the endemic patterns of violence within them, share key features with sites of captivity and confinement of nonhuman animals – in sites of execution and slaughter, in sites of research testing, and in sites of exploited entertainment and commodified labor. Within these carceral sites I have identified a set of 'carceral logics' that underlie their processes and profits, including that of animalization, racialization, and criminalization of certain vulnerable populations, showing how these carceral logics are foundational to their continued operation. Dayan (2016: 3) evocatively captures in just a few lines the material, social, representational, and geographical proximity of this vulnerability:

> City streets are sometimes dangerous – if you happen to be poor, black, or a pit bull. Rural areas are sometimes dangerous – if you happen to be poor, white, or a pit bull. This catalog puts in relation humans and dogs in what might seem like a tasteless or even racist merging. But the taint of proximity matters. ... The pit bull – targeted and reviled, unless saved by a middle-class urban white – joins whites and blacks in a jagged and unlikely bond.

While we might recognize almost infinite differences across the various human and nonhuman groups discussed in the preceding chapters, as I have repeatedly noted, these differences should not prevent us from acknowledging the entangled structures, forms, operations, practices, and embodied experiences that span species boundaries at these sites.

In this Afterword I take up from Chapter 5 questions related to prisoner and animal rights movements, framing them within the limits of the legal and political frameworks touched upon in the preceding chapters. These, for the most part, remain intractably connected to welfarist/reformist and abolitionist prisoner and animal liberation movements that tend to be based upon ultimately unstable human–animal divisions. Acknowledging the limitations of hierarchical, taxonomic, humanist, and speciesist carceral logics is fundamental to moving forward in developing what Kim (2015) and Braverman (2015) put forward as meaningful 'multi-optic' post-humanist ways of thinking needed to challenge the everyday norms of violence intrinsic to industrial production in the United States today. I also continue my discussion of the potential for a non-anthropocentric ethics in seeking justice for incarcerated humans and captive animals. Many approaches might take us forward, but I highlight those advocating for the potential of the affective and visceral (see Woodward and Lea 2010).

Both carceral geography and critical animal studies chart out many challenges and offer many insights well beyond the reaches of this book. I sought to place into conversation a rather specific band of scholarship and activism on the epistemologies of industrial violence, and in so doing have left to others many important and pervasive issues surrounding mass incarceration and mass exploitation and killing of nonhuman animals that other scholars, such as post-humanist feminist scholars, have most fruitfully taken up (e.g. Gaard 2012; Hovorka 2015). That the primary sites of institutionalized violence that I discuss occur in male-dominated industries should not go unnoticed, for example (Deckha 2010: 29). Discussions and debates surrounding 'Afro-pessimism' resonate with many of the issues I raise as well. I agree with McKittrick's (2011: 961N6) analysis, for example, that works such as mine carry the risk of reinforcing a stereotypical representation of 'Black geographies' as places of pathology, lifelessness, risk, fear, health disparities, and so on – indeed, as equatable to zones of violence, death, and displacement themselves. As McKittrick rightly observes, "these aren't wrong, but they articulate that Black geographies are, from the outset, the lowest of the low and lifeless." Suffice it to say, critical resistance to the Prison Industrial Complex as we know it today is occurring on many levels and fronts, and I hope that my part in exposing some of the interlocking processes involved will aid in that work.

Reform or abolition: limits to a trans-species rights movement

When considering ways to challenge prisoner and animal vulnerability, exploitation, and killing in the carceral spaces outlined in this book, it is helpful to return to Acampora's (2006: 109) observation that progressives' call for the reform or improvement of the carceral spaces of the zoo and the prison are markedly similar. Generally speaking, both the prisoner and animal rights movements have tended to focus either on reform of those institutions (the 'welfarist' approach) or eradication of them (the 'abolitionist' approach). Such movements map onto similar historical trajectories, particularly that of the early 20th-century Progressive Era and the anti-vivisection movement, and the emergence of Civil Rights and other social justice and political action work in the 1960s and 1970s. Current and diffuse responses to the hyper-industrialization of prisons, agriculture, and medicine typically have entrenched the welfarist vs. abolition approaches, with scholars now suggesting ways to move beyond them (e.g. Braverman 2015; Kim 2015).

As defined by protections provided within the legal arena, it is at least arguable that the welfarist movement(s) have been the single most effective means of challenging abusive treatment of captive animals (Finsen and Finsen 1994; Bears 2006). Is it not better to have more space, ample and good food, medical care, and cleaner surroundings, than not? Such reforms do enhance the subject's dignity (Gruen 2014b). And in the absence of such protections, abuse within these carceral spaces would be undoubtedly worse. Yet such reforms also distract from the core problems; as Francione (2008) would argue, they aim to 'reform the way we exploit, not the exploitation itself'. In the case of animals, reformist approaches re-entrench their instrumentalist and use value to humans; and in the case of prisoners – if reforms had indeed taken place in the past few decades, which they have not (James 2005; Alexander 2012) – they would sidestep the more difficult questions of why the Prison Industrial Complex and its racialized underpinnings proliferate in the first place. And in the cases of both humans and animals, legal regimes, as I have discussed throughout, are fundamentally anthropocentric and racialized – protecting only some subjects (Cacho 2014; Deckha 2013a).

The scale of any 'progress' made also seems negligible in comparison to the enormous lobbying power yielded by various stakeholders in the Medical-, Agricultural-, and Prison Industrial Complexes. I have addressed the limitations of using U.S. legal frameworks as a litmus test for rights – of 'humane' treatment and so forth – throughout. With animals, arguments for their protections due to their similarity with humans leave the vast majority of them as an exploitable ('animalized') resource for human consumption (Cochrane 2013; Wise 2004; Braverman 2013b; 2015). Such discussions also rest on the political agency, citizenship status, and personhood status of human and nonhuman individuals (Cacho 2014; Deckha 2010; Francione 2004).

The animal abolition movement, or 'new abolition' movement, seems to view veganism as the only means of liberating nonhuman animals from human tyranny (Francione 2000; 2008; Best 2014; Steiner 2008; 2013; and many others). "Abolition" as a descriptor has been borrowed (appropriated?) from the movement to abolish chattel slavery in the 19th century, and now finds traction within both the animal and prison abolition movements. Thus, the very use of the term 'abolition' posits a sort of parallel between the treatment of nonhuman animals and that of Black people under the racialized slave system; and as some would have it, that animals are 'enslaved' in the manner, or worse, than human slaves. Wise (2004: 20), for example, asserts that the

> interdependence of our society, enmeshed in the use of nonhuman animals, dwarfs nineteenth-century slave society. Then it was possible to avoid complicity. Today nonhuman animal products are so omnipresent that one cannot live and not support the abuse of nonhuman animals.

One can appreciate the passion evoked in Wise's important assertion – that we are all complicit in wholesale violence against animals. His position is nonetheless debatable, not least on the grounds that it has been well documented that the integration and growth of the U.S. (and indeed, world economy) was based on southern plantation cotton, tobacco, rice, and sugar. As Frederick Douglass declared, 'the whole nation was complicit in it' and benefited from it (as quoted Zinn 1999: 182–191). But even if that were not the case, the metaphor of animal enslavement – that animals are the 'new slaves' – is to many a problematical co-optation and appropriation of a term and concept that *belongs* to human (racial) slavery and cannot be separated from the transatlantic human slave trade.

There is something provocative about the abolition movement(s) that help to both define the carceral and connect prisoners and animals to it – "abolition" itself as resistance to carceral logics. Yet again, there is a problematical slippery slope when applying the slavery–abolition dyad to other historical, geographical, or social contexts, if for no better reason than that human/Black slavery is not over. This is evidenced in myriad ways but perhaps most pointedly within the 'prison plantation' where abolition struggles continue (Franklin 1989; James 2005; Gilmore 2007; Davis 2003; DuVernay 2016; see Chapter 4). Hart (2014) points out the problematical appropriation and ethical analogy of the slavery concept for unborn fetuses in the abortion debate. To Sexton, this type of generalizing leads inexorably to the 'all lives matter' refrain, which diminishes the import of the historical, literal enslavement of Black people. Instead, as Sexton (2016), Ko (2016), and Jackson (2013) argue, the struggle for both human and animal liberation depends upon challenging and "deposing the human"; i.e. recognizing that the logic of humanness itself can be deposed by contesting hierarchies and logics that privilege putatively 'ideal' humans and renders others 'animal'. Part of this work is in recognizing that *we are all animals*. Thus the category 'human' itself, rather than being reified, should be challenged.

Many critical animal studies scholars debate whether it is hierarchical 'speciesist' thinking itself that reinforces the human in this way, and fosters dominations (e.g. Socha 2013: 223–229; Hudson 2011: 1661). Is the problem one of 'species' itself, and the need to rethink and undo what we might take as natural or cultural categories of difference? Legal scholars such as Braverman (2013a; 2013b; 2015) are examining the incoherence of animal classification systems and the limits, overlaps, and deep ambiguities of the species concept. Picking up from Chapter 3, Kim (2010: 70) challenges such taxonomies which lead, inevitably, to dualistic, hierarchical, and supremacist thinking. Her way forward, again, is in developing a *multi-optic vision* that allows us to simultaneously view the optic of animal oppression and struggle alongside the optic of racial oppression and struggle (Kim 2015: 19–20). 'Optic' here refers to standing within and among, as opposed to panoptical viewing from 'above'. Kim suggests

> a way of seeing that takes disparate justice claims seriously without privileging any one presumptively … Multi-optic seeing entails seeing *from within* various perspectives, moving from one vantage point to another, inhabiting them in

turn, holding them in the mind's eye at once …. [allowing us] to grapple with the existence and interconnectedness of multiple group experiences of oppression. Multi-optic vision encourages a reorientation toward an *ethics of mutual avowal*, or open and active acknowledgement of connection with other struggles.

One important step in attempting to see the interconnection of oppressions and struggles must be to examine the various ontological statuses of those involved in various struggles, challenging the various classifications, categorizations, and hierarchies among races and species, which will ultimately allow us, as Deckha (2010) and Cacho (2014) further assert, to foundationally challenge the boundary between the human and the animal. Moreover, by illuminating the interconnectedness of oppressions we would be able to build coalitions to challenge the carceral logics that reinforce the structural locations of various beings who are vulnerable, unprotected, and exploited within carceral spaces. And though I have not explicitly addressed this here, it is worth noting the potential intersectionalities of movements – the animal rights movements, environmental movements, workers' rights movements, and prisoner rights movements (see Joyce et al., 2015; Emel et al., 2015; Nast 2015). Emel et al. (2015), for example, propose the intersectionality of 'working' animal rights with those of other (human) workers. Of course, such coalition building can also be limited by the sometimes opportunistic agendas and differing social politics and agendas of various actors (Bonds 2015 offers a useful case from within Wisconsin's prison industry). As noted in Chapter 2, to take one timely example, in the state of Nebraska, liberal and conservative legislators across the political and social spectrum built a coalition to override the governor's veto of a bill that abolished the death penalty – on the various grounds that capital punishment is fiscally inefficient and expensive, ineffective in controlling capital crimes, and out of line with Christian and Catholic moral values (Bosman 2015a). Unfortunately, the governor was later able to secure enough signatures to place the issue as a referendum on the November 2016 ballot, and the referendum passed.

A non-anthropomorphic ethics?

Coming to terms with the ethical questions pertinent to abuses within prisoner and animal carceral spaces can sink us into intractable and unhelpful comparisons, or they can offer opportunities for productive dialogue. With respect to the former, I agree with those who argue that it is unhelpful to base one's ethical position on the cognitive or emotional capacities of the respective human and nonhuman populations abused within these spaces since the vast majority of them will not be recognizably 'like us' (cf. DeWaal 2016; Wise 2016). Of the many pitfalls and limitations of such comparisons is the 'capacity for wrong doing' argument. That is, since animals are wholly innocent creatures, their confinement and captivity within carceral space cannot be justified, whereas some humans, due to their potential for evil and wrong-doing, might belong there. Without rehashing my argument about carceral logics that sustain the Prison Industrial Complex, it seems that such a position unhelpfully reasserts a duality, a dividing line and boundary between humanness and animalness, especially a foundational 'criminal-animal' nature of some (racialized) humans, along with an innocent animal 'nature' of some nonhumans (Cacho 2014).

How can we develop a trans-species, non-anthropomorphic ethics that engages the progressive political agendas of carceral geography and critical animal studies? What would such an ethics do, and how could such an ethics be effective? Many geographers are thinking about environmental ethics itself through animal studies. Philo and Wilbert (2000) for example, established a political agenda for human–animal geographical co-existence that sought to respect the inherent territoriality of animal lives, refraining from binding animals rigidly to our own spatial orderings, to 'grant them more room'. As Gillespie and Collard (2015: 9) provocatively suggest, 'multispecies justice' might be found more in distance rather than proximity and intimacy, in difference more than similarity (Chapter 4).

I have explicitly engaged the works of philosophers such as Gruen (2011; 2014a), Guenther (2013), and Francione (2004; 2008) throughout, but it is probably safe to assume that carceral geographers and critical animal scholars generally, like them, choose the work they do because they share an ethical commitment that seeks to challenge the atrocities that take place daily within the carceral spaces discussed in this book. Among many of the challenges faced are those to do with the choice of our research methods. Groling (2014), for example, addresses some of the ethical constraints of ethnographic research in the slaughterhouse or experimental lab, questioning how scholars can maintain their positionality as 'liberationists' as well as bona fide scholars if their (our) interest in empirical research involves going undercover to reveal abuses and atrocities – i.e. that involve actions that we fundamentally oppose (also see Buller 2015). This adds to the methodological complexities associated with research on vulnerable populations that I outlined in the Introduction, which are further challenged by laws that attempt to prohibit or punish such undercover work in the first place (e.g. Kyle and Sewell 2015).

Of the carceral sites I examine in this book, some of the most mainstream and publicized ethical (and bioethical) questions raised are with respect to laboratory experimentation on nonhuman animals. Greenhough and Roe (2010: 43–44; 2011), specifically addressing animals used in scientific research, draw on Haraway's (2008) notion of ethics as performance to advocate for a 'relational ethics' that seeks to share suffering with others. "Shared suffering" entails allowing our imaginations to "articulate, feel, be open and receptive to the suffering which emerges as a result of the experimental process." Shared suffering emphasizes embodied co-presence, such as that of the animal caretaker and caged guinea pigs. But it also requires an acknowledgment that one's way of life depends on the suffering of others, which can become palatable when one "seek[s] less painful practices and ways of being." As Greenhough and Roe (2010: 44) add, ethical relating and caring "at a distance" typifies the expert forum of the conventional research ethics committee, which effectively excludes all those who are unable to articulate their own views in a 'rational, human voice' – animals, babies and children, prisoners, those with mental disorders, etc. – and who must be represented by proxy. The pain and suffering of these individuals might be abstractly acknowledged, but it is not felt or shared in an embodied way (also see Greenhough and Roe 2011).

Such an approach has the potential to accomodate the suffering, if you will, rather than attempting to eliminate it – not unlike the ways that nurses in the Nazi camps described their work during the Nuremberg trials. Lagerwey (2003) and Benedict (2003) describe the main coping mechanism employed by these nurses, which was in considering their work as "caring" for the doomed camp victims. Tuan (1999: 110) calls such compartmentalizations "disassociative monstrosities" – the mental boundaries humans construct around abuse if physical ones are not possible. To quote Tuan,

> may not the ease with which we can, between bouts of laughter and chatting, cut open a turkey and pluck out its innards, prepare us to commit comparable outrage against human beings we consider less than fully human, when circumstances permit?

Among the more well-known examples of attempting to experience the carceral suffering of others is Temple Grandin's 'ethical slaughterhouse' (2006; 2012). As noted in previous chapters, Grandin, also via an embodied somatic sensibility, sought to design slaughterhouses that reduced the stress of animals passing through them – rather than engage directly with the ethics of killing animals for human consumption in the first place. Many (including Greenhough and Roe 2010) acknowledge Grandin's taking seriously an embodied practice of sharing suffering to 'actively learn from animals'; yet the outcome seems to be simply having found an easier way to discipline and kill them. I find the somatic sensibility proposed by Dayan (2016) more compelling. Dayan argues against activism focused on 'animal rights' because these subject to "suspicion and penalty any animal of vigor, independence, intelligence, and yes, capacity for danger" (Waxman 2016). Dayan seeks an approach 'outside of reason', that can be felt but not always understood, perceived but not comprehended (2016: xvi; also see Guenther 2013). Her 'becoming dog' refers to inhabiting and privileging spaces of the primitive and pre-linguistic, of knowing animals beyond or outside of cognition; in spaces of feeling but not of sentiment.

Grandin aside, critical justice work based on affect and viscerality seems to offer one of the most productive ways forward. Recall Hemsworth (2015) alerting us of how important sound is within the prison environment (Chapter 2). Smith (2002: 49) proposes that in their final moments of life, animal "voices can awaken us from our ethical apathy … [so] how can we fail to hear and respond to the Other's voice in a time of need?" In Chapter 2 I discussed the limits of optics and visuality – 'if slaughterhouses had glass walls, we'd all be vegetarians' (Pollan 2002; Rasmussen 2015: 54). Pachirat (2011; 2016) teaches, though, that the 'frontiers of repugnance' are expanding at many carceral sites, indicating that no particular affective or visceral response can be guaranteed through such (visual) sensory exposure. Rasmussen (2015: 56), too, troubles the assumption that 'seeing' something violent or troubling transforms peoples' ethical sensibilities or leads to particular forms of political action. In Rasmussen's examples, violence towards animals is aestheticized and eroticized, such that the harm actually mobilizes desire for sexual gratification within the viewer rather than producing disgust. But as Rasmussen (2015: 66) also usefully points out, there is always a broader set of power relations and representational politics at work to produce particular affective responses. To understand the complex politics of visibility or any sensory reaction we must consider the ways in which affect is not just an individual's 'private possession', but rather a "socially operated inscription of who is sufficiently like us in the ways that guide deliberations." This, she astutely proposes, is a *political* rather than an ethical response to violence.

Such insights have been invaluable to me personally, especially via my involvement with a local nonprofit prisoner rights group, the Lewisburg Prison Project. I had always thought that if only the public was educated to *know* what was happening on the inside of prisons within our district, surely their compassion and outrage would be aroused and ameliorative action would follow (Morin 2013). I assumed that particular responses would follow from incontrovertibly 'knowing' something to be the case. But while there are different kinds of knowing, ways of knowing, and paths to knowing (e.g. rational and intellectual versus emotional and visceral), it is clear that no particular outcome can ever be assured from 'knowing'. And yet it is also clear that accurate information and education are important – there is a great deal of ignorance and/or deliberate misinformation in circulation that must be dispelled. One of the greatest misconceptions about prisons and jails today, for example, is that the violence occurring within their walls originates solely in the individual; that prisoners are locked up because they are bad people, driven to crime and trouble from some indelible social or psychological cause, and that their 'criminal' nature will follow them wherever they go. This, as opposed to the recognition that violence originates from the inherent structure and operation of these spaces (Morin 2016), and because incarcerated people often lack opportunities and resources that others take for granted.

I am reminded of Dinesh Wadiwel's *The War Against Animals* (2015) that provocatively argues that it is the human "luxury of stupidity" that allows our active disavowal of the extreme forms and extent of violence we cause to animals. Wadiwel argues that epistemic injustice requires mechanisms that "willfully withhold hermeneutical resources" that are necessary to make meaningful decisions. Whether we frame the disavowal as stemming from ignorance (which to me would imply being 'uninformed') or stupidity (which implies either a lack of intelligence needed to become informed or a willful decision to avoid it), it is important to keep in mind that in any case there is no 'universal' knowing. Individual and group habits, imaginations, knowledge forms, and so on will always apply differently to different populations because they occupy different social locations. That may seem to lead right back to where I started, with an acknowledgement that there are no guarantees that 'visceral knowing' will intersect with different populations in any sort of guaranteed or predictable way.

Hayes-Conroy and Martin (2010) offer a useful intervention with which I will close. In their study of how individuals become actively involved in the Slow Food Movement, they show how important the roles of emotion, feeling, and sensation are in developing political activism. One of the most important insights they offer is that the 'awakening and training of the senses as well as the intellect' is possible through intentional education (2010: 276). Their evidence suggests that it is possible, through activities and educational events, as well as visceral training (in their case, tasting food), that an appreciation for particular kinds of foods and thus the political meaning behind them and needed to produce them, can be cultivated. Perhaps we need to take a lesson from these scholars' playbook, even if community building and political action around slow food and sustainable local food production are infinitely more pleasurable and 'palatable' than those surrounding the shifts in lifestyle and perspective that would be required for prisoner and animal justice (e.g. veganism; questioning the illusion of safe streets that prisons putatively provide). These authors recognize, though, that the visceral is experiential, it is bio-social, it is relational, it is developmental, and it is heterogeneous (2010: 272). Opportunities for the 'power to act' affectively and viscerally (Woodward and Lea 2010: 161) on behalf of the most vulnerable among us are indeed multiple and varied. Carceral geographers and critical animal scholars and activists have an important – essential – role to play in exposing and combatting the everyday norms of violence intrinsic to industrial production today.

References

Abadie, R. (2010) *The Professional Guinea Pig: Big Pharma and the Risky World of Human Subjects*. Durham NC: Duke University Press.

Acampora, R. (2005) Zoos and eyes: Contesting captivity and seeking successor practices. *Society and Animals* 13: 69–86.

Acampora, R. (2006) *Corporeal Compassion: Animal Ethics and Philosophy of Body*. Pittsburgh PA: University of Pittsburgh Press.

Acampora, R. (ed) (2010) *Metamorphoses of the Zoo: Animal Encounter after Noah*. New York: Lexington Books.

ACLU (American Civil Liberties Union) (2017) National prison project, prisoners' rights. Available at: https://www.aclu.org/prisoners-rights.

Agamben, G. (1998) *Homo Sacer: Sovereign Power and Bare Life*. Stanford CA: Stanford University Press.

Agamben, G. (2004) *The Open: Man and Animal*. Stanford CA: Stanford University Press.

Agamben, G. (2005) *State of Exception* (trans by K. Attell). Chicago: University of Chicago Press.

Alexander, M. (2012) *The New Jim Crow: Mass Incarceration in the Age of Colorblindness*. New York: New Press.

Anonymous (2016) Personal interview with author, March.

Anonymous Pennsylvania prisoner (2017) Correspondence to Lewisburg Prison Project, 20 July.

Appadurai, A. (1988) Introduction: Place and voice in anthropological theory. *Cultural Anthropology* 3(1): 16–20.

Armstrong, S.J. and Botzler, R.G. (2003) *The Animal Ethics Reader*. New York: Routledge.

Arrigo, B.A. and Bullock, J.L. (2008) The psychological effects of solitary confinement on prisoners in supermax units. *International Journal of Offender Therapy and Comparative Criminology* 52(6): 622–640.

Bales, W.D. and Miller, C.H. (2012) The impact of determinate sentencing on prisoner misconduct. *Journal of Criminal Justice* 40: 394–403.

Barad, K. (2003) Posthumanist performativity: Towards an understanding of how matter comes to matter. *Signs* 28(3): 801–831.

Barnes, T. (2005) Audit says prison work program poorly run. *Pittsburgh Post-Gazette*, 14 September.

Basu, M. (2013) Death row diary offers a rare glimpse into a morbid world. *CNN News*, 18 June.

Basu, M. (2015) Botched Oklahoma execution haunts inmate as death nears. *CNN News*, 15 January.

Bears, D. (2006) *For the Prevention of Cruelty: The History and Legacy of Animal Rights Activism in the United States*. Athens OH: Swallow Press/Ohio University Press.

Beardsworth, A. and Bryman, A. (2001) The wild animal in late modernity: The case of Disneyization of zoos. *Tourist Studies* 1: 83–104.

Beirne, P. (1994) The law is an ass: Reading E.P. Evans' *The Medieval Prosecution and Capital Punishment of Animals*. *Society and Animals* 2(1): 27–46.

Beirne, P. (2009) *Confronting Animal Abuse: Law, Criminology, and Human-Animal Relationships*. Lanham MD: Rowman and Littlefield.

Beirne, P. (2011) A note on the facticity of animal trials in early modern Britain: Or, the curious prosecution of farmer Carter's dog for murder. *Crime Law Soc Change* 55: 359–374.

Benbow, S.M.P. (2000) Zoos: Public places to view private lives. *The Journal of Popular Culture* 33: 13–23.

Benbow, S.M.P. (2004) Death and dying at the zoo. *The Journal of Popular Culture* 37(3): 379–398.

Benedict, S. (2003) Caring while killing: Nursing in the 'euthanasia' centers. In Baer, E. and Goldenberg, M. (eds) *Experience and Expression: Women, the Nazis, and the Holocaust*. Detroit MI: Wayne State University Press, 95–110.

Bentham, J. (1781) *An Introduction to the Principles of Morals and Legislation*. Oxford: Claredon Press.

Berger, J. (1980) *About Looking*. New York: Pantheon.

Best, S. (2014) The new abolitionism: Capitalism, slavery and animal liberation. In Best, S. (ed) *The Politics of Total Liberation*. New York: Palgrave Macmillan, 21–49.

Best, S., Kahn, R., Nocella, A.J.II, and McLaren, P. (eds) (2011) *The Global Industrial Complex: Systems of Domination*. New York: Lexington Books.

Bichell, R. (2016) A fix for gender-bias in animal research could help humans. *All Things Considered*, National Public Radio, 10 February.

Big Pharma (2016) Available at: www.drugwatch.com/manufacturer/.

Bilefsky, D. (2015) Zoo's public dissection of lion makes Denmark again a target of outrage. *New York Times*, 15 October.

Birke, L. (2003) Who – or what – are the rats (and mice) in the laboratory? *Society and Animals* 11(3): 210–213.

Bittman, M. (2011) Who protects the animals? *New York Times*, 26 April.

Blackfish (2013) G.C. Thwaite, Director. Magnolia Pictures and CNN Films.

Blinder, A. (2015) Mississippi cuts work program for prisoners. *New York Times*, 3 June.

Blue, G. and Alexander, S. (2015) Coyotes in the city: Gastro-ethical encounters in a more-than-human world. In Gillespie, K. and Collard, R-C. (eds) *Critical Animal Geographies: Politics, Intersections and Hierarchies in a Multi-species World*. London: Routledge, 149–163.

Bonds, A. (2009) Discipline and devolution: Constructions of poverty, race and criminality in the politics of rural prison development. *Antipode* 41(3): 416–438.

Bonds, A. (2015) From private to public: Examining the political economy of Wisconsin's private prison experiment. In Morin, K.M. and Moran, D. (eds) *Historical Geographies of Prisons: Unlocking the Usable Carceral Past*. London: Routledge, 205–218.

Bosman, J. (2015a) Nebraska bans death penalty, defying a veto. *New York Times*, 27 May.

Bosman, J. (2015b) Nebraska to vote on abolishing death penalty after petition drive succeeds. *New York Times*, 16 October.

Bostock, S. (1993) *Zoos and Animal Rights: The Ethics of Keeping Animals*. London: Routledge.

Bosworth, M. (2010) Understanding life in immigration detention. Presentation at the British Society of Criminology Conference, University of Leicester, 13 July.

Bosworth, M. and Kaufman, E. (2011) Foreigners in a carceral age: Immigration and imprisonment in the United States. *Stanford Law and Policy Review* 22: 429–454.

Braverman, I. (2011) States of exemption: The legal and animal geographies of American zoos. *Environment and Planning A* 43: 1693–1706.

Braverman, I. (2013a) *Zooland: The Institution of Captivity*. Stanford CA: Stanford University Press.

Braverman, I. (2013b) Animal mobilegalities: The regulation of animal movement in the American City. *Humanimalia* 5(1): 104–135.

Braverman, I. (2014) Captive for life: Conserving extinct in the wild species through ex situ breeding. In Gruen, L. (ed) *The Ethics of Captivity*. New York: Oxford University Press, 193–212.

Braverman, I. (2015) More-than-human legalities: Advocating an 'animal turn' in law and society. In Sarat, A. and Ewick, P. (eds) *The Handbook of Law and Society*. Chichester: John Wiley & Sons, 307–321.

Bronx Zoo: Experiences (2017) Available at: http://bronxzoo.com/activities.

Brower, C.H. (2004) The lives of animals, the lives of prisoners, and the revelations of Abu Ghraib. *Journal of International Law* 37: 1353–1388.

Bruggeman, S.C. (2012) Reforming the carceral past: Eastern State Penitentiary and the challenge of the twenty-first-century prison museum. *Radical History Review* 113: 171–186.

Buckley, A. (2016) Ava DuVernay on modern slavery in America. *New York Times*, 5 October.

Bulbeck, C. (2010) Respectful stewardship of a hybrid nature: The role of concrete encounters. In Acampora, R. (ed) *Metamorphoses of the Zoo: Animal Encounter After Noah*. New York: Lexington Books, 83–102.

Buller, H. (2014) Animal geographies I. *Progress in Human Geography* 38(2): 308–318.

Buller, H. (2015) Animal geographies II: Methods. *Progress in Human Geography* 39(3): 374–384.

Burton, T.G. (1971) The hanging of Mary, a circus elephant. *Tennessee Folklore Society* 37: 1–8.

Butler, J. (2004) *Precarious Life: The Powers of Mourning and Violence*. New York: Verso.

Butler, J. (2009) *Frames of War: When Is Life Grievable?* New York: Verso.

Cabana, D.A. (1996) *Death at Midnight: The Confession of an Executioner*. Boston: Northeastern University Press.

Cacho, L.M. (2014) *Social Death: Racialized Rightlessness and the Criminalization of the Unprotected*. New York: New York University Press.

Card, C. (2003) Genocide and social death. *Hypatia* 18(1): 63–79.

Carter, B. and Charles, N. (2011) Human–animal connections: An introduction. In Carter, B. and Charles, N. (eds) *Human and Other Animals: Critical Perspectives*. London: Palgrave Macmillan, 1–27.

Carvajal, D. (2014) On display, and on a hot seat: 'Exhibit B', a work about human zoos. *New York Times*, 25 November.

Casal, P. (2003) Is multiculturalism bad for animals? *The Journal of Political Philosophy* 11(1): 1–22.

Castree, N. (2003) Commodifying what nature? *Progress in Human Geography* 27(3): 273–297.

Castree, N. (2004) The geographical lives of commodities: Problems of analysis and critique. *Social and Cultural Geography* 5(1): 21–35.

Castree, N. and Sparke, M. (2000) Introduction: Professional geography and the corporatization of the university: Experiences, evaluations, and engagements. *Antipode* 32(3): 222–229.

Che, D. (2005) Constructing a prison in the forest: Conflicts over nature, paradise and identity. *Annals of the Association of American Geographers* 95(4): 809–831.

Christianson, S. (2010) *The Last Gasp: The Rise and Fall of the American Gas Chamber.* Berkeley: University of California Press.

Chrulew, M. (2010) From zoo to zoopolis: Effectively enacting Eden. In Acampora, R. (ed) *Metamorphoses of the Zoo: Animal Encounter After Noah.* New York: Lexington Books, 193–220.

Clark, J.L. (2014) Labourers or lab tools? Rethinking the role of lab animals in clinical trials. In Taylor, N. and Twine, R. (eds) *The Rise of Critical Animal Studies: From the Margins to the Centre.* London: Routledge, 140–164.

Cloyes, K.G., Lovell, D., Allen, D.G., and Rhodes, L.A. (2006) Assessment of psychosocial impairment in a supermaximum security unit sample. *Criminal Justice and Behavior* 33(6): 760–781.

Cobb, J. (2017a) The banal horror of Arkansas's executions. *The New Yorker*, 5 April.

Cobb, J. (2017b) Old questions but no new answers in the Philando Castile verdict. *The New Yorker*, 22 June.

Cochrane, A. (2013) Cosmozoopolis: The case against group-differentiated animal rights. *Law, Ethics and Philosophy* 1. Available at: www.raco.cat/index.php/LEAP/article/view/294785.

Coetzee, J.M. (1999) *The Lives of Animals.* Princeton NJ: Princeton University Press.

Cohen, C. (1979) Medical experimentation on prisoners. In Robinson, W. and Prichard, M. (eds) *Medical Responsibility: Paternalism, Informed Consent, and Euthanasia.* Clifton NJ: Humana Press, 57–74.

Collard, R-C. (2012) Cougar–Human entanglements and the biopolitical un/making of safe space. *Environment and Planning D: Society and Space* 30(1): 23–42.

Collard, R-C. (2014) Putting animals back together, taking commodities apart. *Annals of the Association of American Geographers* 104(1): 151–165.

Collard, R-C. and Dempsey, J. (2013) Life for sale? The politics of lively commodities. *Environment and Planning A* 45(11): 2682–2699.

Collard, R-C. and Gillespie, K. (2015) Introduction. In Gillespie, K. and Collard, R-C. (eds) *Critical Animal Geographies: Politics, Intersections, and Hierarchies in a Multispecies World.* London: Routledge, 1–16.

Colombino, A. and Steinkrueger, J. (2015) Methods in human-animal studies: The question(s) of animal(s) in practice. Neue Kulturgeographie XIII conference, University of Graz, Austria, January.

Crete, P. (2017) Punitive healing and penal relics: Indigenous prison labour and the (re)production of cultural artefacts. In Wilson, J.Z., Hodgkinson, S., Piché, J., and Walby, K. (eds) *The Palgrave Handbook of Prison Tourism.* London: Palgrave Macmillan, 969–988.

Crewe, B. (2017) Depth, weight, tightness, breadth: Spatial metaphors and the texture of imprisonment. Paper delivered at the 2nd International Conference for Carceral Geography, University of Birmingham, U.K.

Crewe, B., Warr, J., Bennett, P., and Smith, A. (2014) The emotional geography of prison life. *Theoretical Criminology* 18: 56–74.

Critical Resistance (2017) The prison industrial complex. Available at: http://criticalre sistance.org/about/not-so-common-language.

Critter Cave Column, The (2010). Available at: https://crittercave.wordpress.com.

Cunningham, M.D. and Vigen, M.P. (2002) Death row inmate characteristics, adjustment, and confinement. *Behavioral Sciences and the Law* 20: 191–210.

Czajka, A. (2005) Inclusive exclusion: Citizenship and the American prisoner and prison. *Studies in Political Economy* 76: 111–142.

Davies, G. (2013) Mobilizing experimental life: Spaces of becoming with mutant mice. *Theory, Culture and Society* 30(7/8): 129–153.

Davis, A. (2003) *Are Prisons Obsolete?* New York: Seven Stories Press.

Davis, M. (1990) *City of Quartz.* New York: Verso.

Davis, M. (1995) Hell factories in the field. *The Nation* 260: 229–234.

Dayan, C. (2001) Legal slaves and civil bodies. *Nepantla: Views from the South* 2, 1: 1–39.

Dayan, C. (2007) *The Story of Cruel and Unusual.* Cambridge MA: MIT Press.

Dayan, C. (2011) *The Law Is a White Dog: How Legal Rituals Make and Unmake Persons.* Princeton NJ: Princeton University Press.

Dayan, C. (2016) *With Dogs at the Edge of Life.* New York: Columbia University Press.

Death Penalty Information Center (2017) Available at: www.deathpenaltyinfo.org/dea th-row-inmates-state-and-size-death-row-year.

Death Row Diaries (1999) Available at: www.soundportraits.org/deathrow/.

Death Row USA (2015) Available at: www.deathpenaltyinfo.org/.

Deckha, M. (2010) The subhuman as a cultural agent of violence. *Journal for Critical Animal Studies* VIII(3): 28–51.

Deckha, M. (2013a) Initiating a non-anthropocentric jurisprudence: The rule of law and animal vulnerability under a property paradigm. *Alberta Law Review* 50(4): 783–814.

Deckha, M. (2013b) Welfarist and imperial: The contributions of anticruelty laws to civilizational discourse. *American Quarterly* 65(3): 515–548.

DeGrazia, D. (2002) *Animal Rights: A Very Short Introduction.* Oxford: Oxford University Press.

DeMello, M. (2014) Rabbits in captivity. In Gruen, L. (ed) *The Ethics of Captivity.* London: Oxford University Press, 77–89.

Derrida, J. (1990) The force of law: The mystical foundation of authority. *Cardozo Law Review* 11: 919.

Derrida, J. (2002) The animal that therefore I am (more to follow). *Critical Inquiry* 28(2): 369–418.

DeWaal, F. (2016) What I learned from tickling apes. *New York Times*, 9 April.

Dinzelbacher, P. (2002) Animal trials: A multidisciplinary approach. *Journal of Inter-disciplinary History* 32(3): 405–421.

Dober, G. (2008) Cheaper than chimpanzees: Expanding the use of prisoners in medical experiments. *Prison Legal News* 15 (March). Available at: https://www.prisonlega lnews.org/news/2008/mar/15/cheaper-than-chimpanzees-expanding-the-use-of-p risoners-in-medical-experiments/.

Dober, G. (2016) *Prison Legal News.* Available at: https://www.prisonlegalnews.org/ news/author/greg-dober/.

DOC Policy Statement: Inmate Compensation DC-ADM 816 (2008) Commonwealth of Pennsylvania Department of Corrections.

Donahue, J. and Trump, E. (2006) *The Politics of Zoos: Exotic Animals and Their Protectors.* DeKalb: Northern Illinois University Press.

Dow, D., Marcus, J., Moon, M., Tyler, J., and Wiercioch, G. (2004) The extraordinary execution of Billy Vickers, the banality of death, and the demise of post-conviction review. *William and Mary Bill of Rights Journal* 13(2): 521–566.

DuVernay, A. (2016) *13th*. Produced by DuVernay, A., Averick, S., and Barish, H.

Dykstra, R.R. (1968) *The Cattle Towns*. Lincoln: University of Nebraska Press.

Eason, J. (2010) Mapping prison proliferation: Region, rurality, race and disadvantage in prison placement. *Social Science Research* 39(6): 1015–1028.

Ehrenreich, B. and Ehrenreich, J. (1971) *American Health Empire: Power, Profits, and Politics*. New York: Random House.

Elder, G., Wolch, J., and Emel, J. (1998) Le pratique sauvage: Race, place, and the human-animal divide. In Wolch, J. and Emel, J. (eds) *Animal Geographies: Place, Politics, and Identity in Nature-Cultural Borderlands*. London: Verso, 72–90.

Elias, N. (2000) *The Civilizing Process*. Malden MA: Blackwell.

Elk, M. and Sloane, B. (2011) The hidden history of ALEC and prisoner labor. *The Nation*, 1 August.

Emel, J., Johnston, C.L., and Stoddard, E.L. (2015) Livelier livelihoods: Animal and human collaboration on the farm. In Gillespie, K. and Collard, R-C. (eds) *Critical Animal Geographies: Politics, Intersections and Hierarchies in a Multi-species World*. London: Routledge, 164–183.

Emel, J. and Urbanik, J. (2010) Animal geographies: Exploring the spaces and places of human–animal encounters. In DeMello, M. (ed) *Teaching the Animal: Human–Animal Studies Across the Disciplines*. New York: Lantern Books, 202–217.

Emel, J. and Wolch, J. (1998) Witnessing the animal movement. In Wolch, J. and Emel, J. (eds) *Animal Geographies: Place, Politics, and Identity in the Nature–Culture Borderlands*. London: Verso, 1–23.

Emel, J., Wilbert, C., and Wolch, J. (2002) Animal geographies. *Society and Animals* 10(4): 407–412.

Equal Justice USA (2016) Available at: http://ejusa.org/learn/cost/.

Evans, E.P. (1906) *The Criminal Prosecution and Capital Punishment of Animals*. London: Heinemann.

Execution Tapes, The (1998) Sound Portraits.com. Available at: http://soundportraits. org/on-air/execution_tapes/.

Exposing abuse on the factory farm (2015) Editorial, *New York Times*, 8 August.

Fanon, F. (1967) *Black Skin, White Masks*. New York: Grove Press.

Finsen, L. and Finsen, S. (1994) *The Animal Rights Movement in America: From Compassion to Respect*. New York: Twayne.

Foucault, M. (1977) *Discipline and Punish: The Birth of the Prison*. New York: Penguin Books.

Foucault, M. (2007) *Security, Territory, Population: Lectures at the Collège de France, 1977–78*. Ed. Michel Senellart. Basingstoke: Palgrave Macmillan.

Foundation for Biomedical Research (FBR) (2017) Nobel prizes. Available at: https:// fbresearch.org/medical-advances/nobel-prizes/.

FPI (Federal Prison Industry Policy) (2017) Available at: https://www.bop.gov/Pub licInfo/execute/policysearch?todo=query&series=8000.

Francione, G.L. (1995) *Animals, Property, and the Law*. Philadelphia: Temple University Press.

Francione, G.L. (2000) *Introduction to Animal Rights: Your Child or Your Dog?* Philadelphia: Temple University Press.

Francione, G.L. (2004) Animals – Property or Persons? In Sustein, C.R. and Nussbaum, M.C. (eds) *Animal Rights: Current Debates and New Directions*. Oxford: Oxford University Press, 108–142.

Francione, G.L. (2008) *Animals as Persons: Essays on the Abolition of Animal Exploitation*. New York: Columbia University Press.

Franju, G. (1949) Blood of the beasts. Available at: www.criterion.com/films/950-eyes-without-a-face.

Franklin, B. (1989) *Prison Literature in America: The Victim as Criminal and Artist*. New York: Oxford University Press.

Fraser, J. (2000) An American seduction: Portrait of a prison town. *Michigan Quarterly Review* 39(4): 775–795.

Gaard, G. (2012) Feminist animal studies in the U.S.: Bodies matter. *DEP* 20: 14–21.

Garland, D. (2005) Capital punishment and American culture. *Punishment and Society* 7(4): 347–376.

Gates, S. (2013) Tiger attack at Oklahoma zoo leaves worker injured after she sticks her arm inside cage. *The Huffington Post*, 6 October.

Giedion, S. (1948; rpt 2013) *Mechanization Takes Command: A Contribution to Anonymous History*. Minneapolis: University of Minnesota Press.

Gill, N., Conlon, D., Moran, D., and Burridge, A. (2016) Carceral circuitry: New directions in carceral geography. *Progress in Human Geography* (DOI online).

Gillespie, L.K. (2003) *Inside the Death Chamber: Exploring Executions*. Boston: Pearson Education.

Gillespie, K. (2012) Nonhuman animal resistance and the improprieties of live property. In Braverman, I. (ed) *Animals, Biopolitics, Law: Lively Legalities*. New York: Routledge, 117–132.

Gillespie, K. (2014) Sexualized violence and the gendered commodification of the animal body in Pacific Northwest US dairy production. *Gender, Place and Culture* 21(10): 1321–1337.

Gillespie, K. (forthcoming) Placing Angola: Race, animality, and human–animal encounters at the Louisiana State Penitentiary.

Gillespie, K. and Collard, R-C. (eds) (2015) *Critical Animal Geographies: Politics, Intersections, and Hierarchies in a Multispecies World*. London: Routledge.

Gillespie, K. and Lopez, P. (2015) Introducing economies of death. In Lopez, P. and Gillespie, K. (eds) *Economies of Death: Economic Logics of Killable Life and Grievable Death*. London: Routledge, 1–13.

Gilmore, R.W. (2007) *Golden Gulag: Prisons, Surplus, Crisis and Opposition in Globalizing California*. Berkeley: University of California Press.

Girgen, J. (2003) The historical and contemporary prosecution and punishment of animals. *Animal Law* 9: 97–133.

Glick, M.H. (2013) Animal instincts: Race, criminality, and the reversal of the 'human'. *American Quarterly* 65(3): 639–659.

Goode, E. (2013) Prisons rethink isolation, saving money, lives and sanity. *New York Times*, 10 March.

Gordon, A. (2011) Some thoughts on haunting and futurity. *Borderlands* 10(2): 1–21.

Gorman, T. (1997) Back on the chain gang: Why the Eighth Amendment and the history of slavery proscribe the resurgence of the chain gang. *California Law Review* 85(2): 441–478.

Gottschalk, M. (2006) *The Prison and the Gallows: The Politics of Mass Incarceration in America*. Cambridge: Cambridge University Press.

Grabell, M. (2017) Cut to the bone: How a poultry company exploits immigrant laws. *The New Yorker*, 8 May.

Grandin, T. (2006) *Thinking in Pictures: My Life in Pictures*. New York: Vintage Books.

Grandin, T. (2012) Humane slaughter: Tour of beef plant featuring Temple Grandin. Available at: www.grandin.com/humane/rec.slaughter.html.

Greenhough, B. and Roe, E. (2010) From ethical principles to response-able practice. *Environment and Planning D: Society and Space* 28: 43–45.

Greenhough, B. and Roe, E. (2011) Ethics, space, and somantic sensibilities: Comparing relationships between scientific researchers and their human and animal experimental subjects. *Environment and Planning D: Society and Space* 29: 47–66.

Gregory, D. (2006) The black flag: Guantanamo Bay and space of exception. *Geografiska Annaler* 88B(4): 405–427.

Griffin, D.R. (1992) *Animal Minds*. Chicago: University of Chicago Press.

Grim, R. (2016a) Mississippi jails are losing inmates, and local officials are 'devastated' by the loss of revenue. *The Huffington Post*, 14 April.

Grim, R. (2016b) Mississippi prison boss defends repossessing inmates to cover budget shortfall. *The Huffington Post*, 15 April.

Groling, J. (2014) Studying perpetrators of socially-sanctioned violence against animals through the i/eye of the CAS scholar. In Taylor, N. and Twine, R. (eds) *The Rise of Critical Animal Studies: From the Margins to the Center*. London: Routledge, 88–110.

Gruen, L. (2011) *Ethics and Animals: An Introduction*. Cambridge: Cambridge University Press.

Gruen, L. (ed) (2014a) *The Ethics of Captivity*. New York: Oxford University Press.

Gruen, L. (2014b) Dignity, captivity, and an ethics of sight. In Gruen, L. (ed.) *The Ethics of Captivity*. New York: Oxford University Press, 231–247.

Gruen, L. (2015a) *Entangled Empathies*. New York: Lantern Books.

Gruen, L. (2015b) Samuel Dubose, Cecil the Lion and the ethics of avowal. *Al Jazeera America*, 31 July.

Gruen, L. (2016a) Carceral logic and an ethics of avowal. Lecture presented at the Race and Animals Summer Institute, Wesleyan University, Middletown CT, June.

Gruen, L. (2016b) The Cincinnati Zoo's problem wasn't that it killed its gorilla. It's that it's a zoo. *The Washington Post*, 1 June.

Gruen, L. (2017) The Last 1000. Available at: http://last1000chimps.com/.

Gruen, L. (forthcoming) *Carceral Logic and the Paradox of Hope*. Manuscript in preparation.

Guenther, L. (2012) Beyond dehumanization: A post-humanist critique of solitary confinement. *Journal for Critical Animal Studies* 10(2): 47–68.

Guenther, L. (2013) *Solitary Confinement: Social Death and Its Afterlives*. Minneapolis: University of Minnesota Press.

Guthman, J. (2011) Bodies and accumulation: Revisiting labour in the 'production of nature'. *New Political Economy* 16(2): 233–238.

Hallman, B. and Benbow, M. (2006) Naturally cultural: The zoo as cultural landscape. *Canadian Geographer* 50: 256–264.

Hallsworth, S. and Lea, J. (2011) Reconstructing Leviathan: Emerging contours of the security state. *Theoretical Criminology* 15: 141–157.

Haney, C. (2008) A culture of harm: Taming the dynamics of cruelty in supermax prisons. *Criminal Justice and Behavior* 35: 956–984.

Hannon, E. (2015) Utah lawmakers vote for firing squad executions as backup if lethal injection unavailable. *Slate*, 11 March.

Haraway, D. (1989) *Primate Visions: Gender, Race, and Nature in the World of Modern Science*. New York: Routledge.

Haraway, D. (2008) *When Species Meet*. Minneapolis: University of Minnesota Press.

Harcourt, B.E. (2010) Neoliberal penality: A brief genealogy. *Theoretical Criminality* 14: 74–92.

Hart, W.D. (2014) Slaves, fetuses, and animals: Race and ethical rhetoric. *Journal of Religious Ethics* 42(4): 661–690

Harvey, D. (1998) The body as accumulation strategy. *Environment and Planning D: Society and Space* 16: 412–421.

Hayes-Conroy, A. and Martin, D.G. (2010) Mobilising bodies: Visceral identification in the Slow Food Movement. *Transactions of the Institute of British Geographers* 35: 269–281.

Health costs and budgets (2017) The Henry J. Kaiser Family Foundation. Available at: http://kff.org/state-category/health-costs-budgets/prescription-drugs/.

Hemsworth, K. (2015) Carceral acoustemologies: Historical geographies of sound in a Canadian prison. In Morin, K.M. and Moran, D. (eds) *Historical Geographies of Prisons: Unlocking the Usable Carceral Past*. London: Routledge, 17–33.

Higgin, M., Evans, A., and Miele, M. (2011) A good kill: Socio-technical organisations of farm animal slaughter. In Carter, B. and Charles, N. (eds) *Humans and Other Animals: Critical Perspectives*. Basingstoke: Palgrave Macmillan, 173–194.

Hodgetts, T. and Lorimer, J. (2015) Methodologies for animals' geographies: Cultures, communication and genomics. *Cultural Geographies* 22(2): 285–295.

Hornblum, A. (1998) *Acres of Skin: Human Experiments at Holmesburg Prison*. New York: Routledge.

Hornblum, A. (2007) *Sentenced to Science: One Black Man's Story of Imprisonment in America*. University Park PA: Pennsylvania State University Press.

Horse slaughter: Separating fact from fiction (2015) *ASPCA Action*. Spring/Summer: 6–10.

Hovorka, A.J. (2015) The Gender, Place and Culture Jan Monk Distinguished Annual Lecture. Feminism and animals: Exploring interspecies relations through intersectionality, performativity and standpoint. *Gender, Place and Culture* 22(1): 1–19.

Hribal, J. (2003) Animals are part of the working class: A challenge to labor history. *Labor History* 44(4): 435–453.

Hudson, L. (2011) A species of thought: Bare life and animal being. *Antipode* 43(5): 1659–1678.

IMS Health (2017) Available at: www.imshealth.com/.

Jackson, Z.I. (2013) Animal: New directions in the theorization of race and post-humanism. *Feminist Studies* 39(3): 669–685.

Jackson, Z.I. (2015) Outer worlds: The persistance of race in movement 'beyond the human'. *Journal of Lesbian and Gay Studies* 21(2–3): 215–218.

James, J. (ed) (2005) *The New Abolitionists: (Neo) Slave Narratives and Contemporary Prison Writings*. Albany: State University of New York Press.

Jamieson, D. (1985) Against zoos. In Singer, P. (ed) *In Defense of Animals*. New York: Harper and Row, 108–117.

Jasper, J.M. (2003) Review: *Animal Rights/Human Rights* by David Nibert. *Contemporary Sociology* 32(3): 344–345.

Jerreat, J. (2014) 'It looks like 21st century slavery': Photographer's images of Arizona's chain gangs evoke a dark period in the country's history. *Daily Mail*, 28 January.

Jones, J. (1981) *Bad Blood: The Tuskegee Syphilis Experiment*. New York: The Free Press.

Jones, M. (2010) 'Impedimenta state': Anatomies of neoliberal penality. *Criminology and Criminal Justice* 10: 393–404.

Jones, M. (2014) Captivity in the context of a sanctuary for formerly farmed animals. In Gruen, L. (ed) *The Ethics of Captivity*. New York: Oxford University Press, 90–101.

Joyce, J., Nevins, J., and Schneiderman, J.S. (2015) Commodification, violence, and the making of workers and ducks at Hudson Valley Foie Gras. In Gillespie, K. and Collard, R-C. (eds) *Critical Animal Geographies: Politics, Intersections, and Hierarchies in a Multispecies World*. London: Routledge, 93–107.

Kane, T. (2017) Federal Bureau of Prisons, Federal Prison Industries Policy. Available at: https://www.bop.gov/PublicInfo/execute/policysearch?todo=query&series=8000#.

Kemmerer, L. (2010) Nooz: Ending zoo exploitation. In Acampora, R. (ed) *Metamorphoses of the Zoo: Animal Encounter After Noah*. New York: Lexington Books, 37–56.

Kim, C.J. (2010) Slaying the beast: Reflections on race, culture and species. *Kalfou* (Inaugural issue, Spring): 57–74.

Kim, C.J. (2011) Moral extensionism or racist exploitation: The use of holocaust and slavery analogies in the animal liberation movement. *New Political Science* 33(3): 311–333.

Kim, C.J. (2015) *Dangerous Crossings: Race, Species, and Nature in a Multicultural Age*. Cambridge: Cambridge University Press.

Kim, C.J. (2016) Lecture presented at the Race and Animals Summer Institute, Wesleyan University, Middletown CT, June.

Kim, C.J. (2017) Murder and mattering in Harambe's house. *Politics and Animals* 2 (Fall): 37–51.

King, B.J. (2013) *How Animals Grieve*. Chicago: University of Chicago Press.

King, K., Steiner, B., and Breach, S. (2008) Violence in the supermax: A self-fulfilling prophecy. *The Prison Journal* 88: 144–168.

Kirksey, S.E. and Helmreich, S. (2010) The emergence of multispecies ethnography. *Cultural Anthropology* 25: 545–576.

Knapp, L. (2017) Death row doctor. *New York Times*, 17 January.

Ko, Syl (2016) Notes from the border of the human–animal divide: Thinking and talking about animal oppression when you're not quite human yourself. *Aphro-ism: Essays on Pop Culture, Feminism, and Black Veganism from Two Sisters*, 13 January. Available at: https://aphro-ism.com/.

Kovensky, J. (2014) It's time to pay prisoners minimum wage. *New Republic*, 15 August.

Kyle, Z. and Sewell, C. (2015) Federal judge strikes down Idaho's 'ag-gag' law. *Idaho Statesman*, 3 August.

Lagerwey, M.D. (2003) The nurses' trial at Hadamar and the ethical implications of health care values. In Baer, E. and Goldenberg, M. (eds) *Experience and Expression: Women, the Nazis, and the Holocaust*. Detroit MI: Wayne State University Press, 111–126.

Landa, A. (2009) *When Medicine and Ethics Meet in the Public Sphere: The Role of Journalism in the History of Bioethics*. Minneapolis: University of Minnesota School of Journalism and Mass Communication.

Law, J. (2012) Notes on fish, ponds and theory. *Norsk Anthropologisk Tidsskrift* 23(3–4): 225–236.

Leder, D. with Greco, V. (2014) Prisoners: 'They're animals' and their animals. In Tysar, S. and Hall, J. (eds) *Philosophy Imprisoned*. New York: Lexington Books, 219–234.

Levasseur, R.L. (2005) Trouble coming every day: ADX – the first year 1996. In James, J. (ed.) *The New Abolitionists: (Neo) Slave Narratives and Contemporary Prison Writings*. Albany: State University of New York Press, 47–55.

Lichtenstein, A. (1996) Chain gang blues. *Dissent* 43(3): 6–10.

Light, D.W. (2017) Demythologizing the high cost of drug research. Available at: www.pharmamyths.net/demythologizing_the_high_costs_of_drug_research_106055.htm.

Loki, R. (2016) The bigger story behind the killing of Cecil the Lion that the media overlooked. Alternet. Available at: www.alternet.org/environment/bigger-story-behind-killing-cecil-lion-media-overlooked.

Lopez, P.J. and Gillespie, K.A. (eds) (2015) *Economies of Death: Economic Logics of Killable Life and Grievable Death*. London: Routledge.

Lorimer, H. (2010) Forces of nature, forms of life: Calibrating ethology and phenomenology. In Anderson, B. and Harrison, P. (eds) *Taking-Place: Non-Representational Theories and Geography*. London: Ashgate, 55–78.

Loyd, J., Mitchelson, M., and Burridge, A. (eds) (2012) *Beyond Walls and Cages: Prisons, Borders, and Global Crisis*. Athens: University of Georgia Press.

Loyd, J. (2012) Race, capitalist crisis, and abolitionist organizing: An interview with Ruth Wilson Gilmore. In Loyd, J., Mitchelson, M., and Burridge, A. (eds) *Beyond Walls and Cages: Prisons, Borders, and Global Crisis*. Athens: University of Georgia Press, 42–54.

Lynch, M. (2000) The Disposal of inmate #85271: Notes on a routine execution. *Studies in Law, Politics, and Society* 20: 3–34.

Magnani, L. and Wray, H. (2006) *Beyond Prisons: A New Interfaith Paradigm for Our Failed Prison System*. Minneapolis: Fortress Press.

Malamud, R. (1998) *Reading Zoos: Representations of Animals and Captivity*. New York: New York University Press.

Malon, D. with S. Hawkins (2007) In Mulvey-Roberts, M. (ed) *Writing for Their Lives: Death Row USA*. Urbana: University of Illinois Press, 131–134.

Maron, D.F. (2014) Should prisoners be used in medical experiments? *Scientific American*, 2 July.

Martel, J. (2006) 'To be, one has to be somewhere'. *British Journal of Criminology* 46(4): 587–612.

Mbembe, A. (2003) Necropolitics. *Public Culture* 15(1): 11–40.

McCormack, S. (2012) Prison labor booms as unemployment rates remain high: Companies reap benefits. *The Huffington Post*, 10 December.

McKee, T.V.S. (2015) The cost of a second chance: Life, death, and redemption among prison inmates and thoroughbred ex-racehorses in Bluegrass Kentucky. In Lopez, P.J. and Gillespie, K.A. (eds) *Economies of Death: Economic Logics of Killable Life and Grievable Death*. London: Routledge, 37–54.

McKittrick, K. (2011) On plantations, prisons, and a black sense of place. *Social and Cultural Geography* 12(8): 947–963.

Mears, D. and Reisig, M. (2006) The theory and practice of supermax prisons. *Punishment and Society* 8: 33–57.

Mele, C. (2017) Ringling Bros. and Barnum and Bailey Circus to end its 146-year run. *New York Times*, 14 January.

Merritt, R.M. and Hurley, S. (2014) Invisible geographies: Violence and oppression in the prison-industrial complex and concentrated animal feeding operations. Paper presented at the 2014 Annual Association of American Geographers Meeting, Tampa, Florida.

MIC (Medical Industrial Complex) (2017) Edu.LearnSoc.org: An insight into human social relations. Available at: http://edu.learnsoc.org/Chapters/21/medical-industrial%20complex.htm.

Miele, M. (2013) Religious slaughter: Promoting a dialogue about the welfare of animals at the time of killing. *Society and Animals* 21: 421–424.

Military Industrial Complex (2017) MilitaryIndustrialComplex.com. Available at: www.militaryindustrialcomplex.com/what-is-the-military-industrial-complex.asp.

Mitford, J. (1973) Experiments behind bars: Doctors, drug companies, and prisoners. *Atlantic Monthly* 76(1): 64–73.

Montford, K.S. (2016) Dehumanized denizens, displayed animals: Prison tourism and the discourse of the zoo. *PhiloSophia* 6(1): 73–91.

Moran, D. (2015a) *Carceral Geography: Spaces and Practices of Incarceration.* Farnham: Ashgate.

Moran, D. (2015b) Budgie smuggling or doing bird? Human–animal interactions in carceral space: Prison(er) animals as abject and subject. *Social and Cultural Geography* 16(6): 634–653.

Moran, D., Gill, N., and Conlon, D. (eds) (2013) *Carceral Spaces: Mobility and Agency in Imprisonment and Migrant Detention.* Farnham: Ashgate.

Moran, D., Turner, J., and Schliehe, A. (2017) Conceptualising the carceral in carceral geography. *Progress in Human Geography* (DOI online).

Morgan, K. and Cole, M. (2011) The discursive representation of nonhuman animals in a culture of denial. In Carter, B. and Charles, N. (eds) *Humans and Other Animals: Critical Perspectives.* New York: Palgrave Macmillan, 112–132.

Morin, K.M. (2008) *Frontiers of Femininity: A New Historical Geography of the Nineteenth-century American West.* Syracuse NY: Syracuse University Press.

Morin, K.M. (2011) *Civic Discipline: Geography in America, 1860–1890.* London: Ashgate.

Morin, K.M. (2013) 'Security here is not safe': Violence, punishment, and space in the contemporary U.S. penitentiary. *Environment and Planning D: Society and Space* 31(3): 381–399.

Morin, K.M. (2015) Wildspace: The cage, the supermax and the zoo. In Gillespie, K. and Collard, R–C. (eds) *Critical Animal Geographies: Politics, Intersections and Hierarchies in a Multi-species World.* London: Routledge, 73–91.

Morin, K.M. (2016) The late-modern American jail: Epistemologies of space and violence. *The Geographical Journal* 182(1): 38–48.

Morin, K.M. (2017) Carceral space: Prisoners and animals. *Antipode: A Radical Journal of Geography* 48(5): 1317–1336.

Morin, K.M. and Moran, D. (2015) *Historical Geographies of Prisons: Unlocking the Usable Carceral Past.* London: Routledge.

Morris, N. (2000) Prisons in the USA: Supermax – the bad and the mad. In Fairweather, L. and McConville, S. (eds) *Prison Architecture: Policy, Design and Experience.* Oxford: Architectural Press, 98–108.

Morton, G. (2008) Prison workers say they're at risk. *The Daily Item*, 8 July.

Moss, M. (2015) U.S. research lab lets livestock suffer in quest for profit: Animal welfare at risk in experiments for meat industry. *New York Times*, 19 January.

Mountz, A. (2011) The Enforcement archipelago: Detention, haunting, and asylum on islands. *Political Geography* 30(3): 118–128.

Mowe, S. (2016) Signs of intelligent life: Carl Safina's evidence that other animals think and feel. *The Sun* 488 (August): 4–11.

Mulvey-Roberts, M. (ed) (2007) *Writing for Their Lives: Death Row USA.* Urbana: University of Illinois Press.

Myers, M.A. (1998) *Race, Labor and Punishment in the New South.* Columbus: Ohio State University Press.

Nance, S. (2013) *Entertaining Elephants: Animal Agency and the Business of the American Circus.* Baltimore MD: Johns Hopkins University Press.

Nast, H.J. (2015) Pit bulls, slavery, and whiteness in the mid- to late-nineteenth century U.S.: Geographical trajectories; primary sources. In Gillespie, K. and Collard, R–C. (eds) *Critical Animal Geographies: Politics, Intersections, and Hierarchies in a Multispecies World.* London: Routledge, 127–145.

National Correctional Industries Association (2015) Available at: www.nationalcia.org/about/top-ten-benefits-of-ci.

National Public Radio (NPR) with The Marshall Project (2016) Inside Lewisburg prison: A choice between a violent cellmate or shackles. 16 October.

Nebraska keeps death penalty (2016) *Lincoln Journal Star*, 10 November.

Netz, R. (2004) *Barbed Wire: An Ecology of Modernity*. Middletown CT: Wesleyan University Press.

Newkirk, P. (2015) *Spectacle: The Astonishing Life of Ota Benga*. New York: Harper Collins.

Nibert, D. (2002) *Animal Rights/Human Rights: Entanglements of Oppression and Liberation*. Lanham MD: Rowman and Littlefield.

Nibert, D. (2003) Humans and other animals: Sociology's moral and intellectual challenge. *The International Journal of Sociology and Social Policy* 23(3): 5–25.

Nibert, D. (2013) *Animal Oppression and Human Violence: Domesecration, Capitalism, and Global Conflict*. New York: Columbia University Press.

NIH (National Institute of Health) (2017) National Center for Advancing Translational Sciences: About new therapeutic uses. Available at: https://ncats.nih.gov/ntu/about.

Nonhuman Rights Project (2015) Available at: www.nonhumanrights.org/.

Norton, J. (2015) Little Siberia, star of the north: The political economy of prison dreams in the Adirondacks. In Morin, K.M. and Moran, D. (eds) *Historical Geographies of Prisons: Unlocking the Usable Carceral Past*. London: Routledge, 168–184.

Orson, D. (2012) 'Million Dollar Blocks' map incarceration's cost. National Public Radio, 2 October.

Orzechowski, K. (2014) Maximum tolerated dose. Available at: http://maximumtolerateddose.org/.

Oshinsky, D.M. (1996) *'Worse Than Slavery': Parchman Farm and the Ordeal of Jim Crow Justice*. New York: The Free Press.

Pachirat, T. (2011) *Every Twelve Seconds: Industrialized Slaughter and the Politics of Sight*. New Haven CT: Yale University Press.

Pachirat, T. (2016) Lecture presented at the Race and Animals Summer Institute, Wesleyan University, Middletown CT, June.

Pacyga, D.A. (2015) *Slaughterhouse: Chicago's Union Stock Yard and the World It Made*. Chicago: University of Chicago Press.

Pelaez, V. (2014) The prison industry in the United States: Big business or a new form of slavery? *Global Research*, 31 March.

Parker, I. (2017) Killing animals at the zoo. *The New Yorker*, January 16.

Patterson, C. (2002) *Eternal Treblinka: Our Treatment of Animals and the Holocaust*. New York: Lantern Books.

Patterson, O. (1982) *Slavery and Social Death: A Comparative Study*. Cambridge MA: Harvard University Press.

Patterson, W.L. (1951) *We Charge Genocide: The Historic Petition to the United Nations for Relief from a Crime of the United States Government Against the Negro People*. New York: International Publishers.

PCI (Pennsylvania Correctional Industries) (2017) Available at: www.cor.pa.gov/PCI/Pages/default.aspx#.WVgS2oqQyCQ.

PCRM (Physicians Committee for Responsible Medicine) (2017) Available at: www.pcrm.org/.

Peck, J. (2003) Geography and public policy: Mapping the penal state. *Progress in Human Geography* 27: 222–232.

Performance Audit of PA Correctional Industries (2005) Available at: www.paauditor.gov/Media/Default/Reports/PCI%20Full%20Audit%20Report.pdf.

Phelps, M. (2011) Rehabilitation in the punitive era: The gap between rhetoric and reality in U.S. prison programs. *Law Soc Rev* 45(1): 33–68.

Philo, C. (1998) Animals, geography, and the city: Notes on inclusions and exclusions. In Wolch, J. and Emel, J. (eds) *Animal Geographies*. London: Verso, 51–71.

Philo, C. and Wilbert, C. (eds) (2000) *Animal Spaces, Beastly Places: New Geographies of Human–Animal Relations*. London: Routledge.

Philo, C. and MacLachlan, I. (2017) The strange case of the missing slaughterhouse geographies. In Wilcox, S. and Rutherford, S. (eds) *Historical Animal Geographies*. London: Routledge (forthcoming).

Pollan, M. (2002) An animal's place. *New York Times Sunday Magazine*, 10 November.

Pratt, J., Brown, D., Brown, M., Hallsworth, S., and Morrison, W. (eds) (2005) *The New Punitiveness: Trends, Theories, Perspectives*. Cullompton, Devon: Willan Publishing.

Pratt, M.L. (1992) *Imperial Eyes: Travel Writing and Transculturation*. London: Routledge.

Prison made products and services, definition of (2003) *Manufacturing in Prison*, Federal Prison Industries Inc. Available at: https://www.bop.gov/policy/progstat/8400_003.pdf.

Raemisch, R. (2014) My night in solitary. *New York Times*, 20 February.

Rasmussen, C. (2015) Pleasure, pain, and place: Ag-gag, crush videos, and animal bodies on display. In Gillespie, K. and Collard, R-C. (eds) *Critical Animal Geographies*. London: Routledge, 54–70.

Raymond, A. and Raymond, S. (1991) *Doing Time: Life Inside the Big House*. Video Verite.

Raymond, J.L. (2009) Ensemble of immigrant actors to perform 'The Story of Our Lives': Personal account of U.S. immigration, the Postville raid and federal detention. *Decorah Journal*, 2 April.

Reiter, K. (2009) Experimentation on prisoners: Persistent dilemmas in rights and regulations. *California Law Review* 97(2): 501–566.

Rhodes, L. (2009) Supermax prisons and the trajectory of exception. *Studies in Law, Politics, and Society* 47: 193–218.

Richards, S. (2008) USP Marion: The first federal supermax. *The Prison Journal* 88: 6–22.

Rifkin, J. (1992) *Beyond Beef: The Rise and Fall of the Cattle Culture*. New York: Dutton.

Rilke, R.M. (trans. by Mitchell, S.) (1982) *Selected Poetry of Rainer Maria Rilke*. New York: Random House.

Robbins, P. (1998) Shrines and butchers: Animals as deities, capital, and meat in contemporary North India. In Wolch, J. and Emel, J. (eds) *Animal Geographies: Place, Politics, and Identity in Nature–Cultural Borderlands*. London: Verso, 218–240.

Rodriguez, D. (2005) *Forced Passages: Imprisoned Radical Intellectuals and the U.S. Prison Regime*. Minneapolis: University of Minnesota Press.

Ross, S. (2014) Captive chimpanzees. In Gruen, L. (ed) *The Ethics of Captivity*. New York: Oxford University Press, 57–76.

Routley, L. (2016) The carceral: Beyond, around, through and within prison walls. *Political Geography* (DOI June).

Rudy, K. (2011) *Loving Animals: Toward a New Animal Advocacy*. Minneapolis: University of Minnesota Press.

Safina, C. (2016) There is someone there. *The Sun* 488(August): 12–13.

Said, E. (1978) *Orientalism: Western Conceptions of the Orient*. London: Kegan Paul.

Samuels, D. (2012) Wild things: Animal nature, human racism, and the future of zoos. *Harper's Magazine*, June.

Schlossberg, T. (2014) Brooklyn man who kicked cat says he shouldn't go to jail. *New York Times*, 1 October.

Schlosser, E. (2001) *Fast Food Nation: The Dark Side of the All-American Meal*. New York: Perennial.

Schrift, M. (2004) The Angola Prison rodeo: Inmate cowboys and institutional tourism. *Ethnology* 43: 331–344.

Senate Bill No. 59 (2017) The General Assembly of Pennsylvania, Prison Industry Enhancement Authority, 12 January.

Sexton, J. (2010) People-of-Colorblindness: Notes on the afterlife of slavery. *Social Text 103* 28(2): 31–56.

Sexton, J. (2016) Presentation at the Race and Animals Summer Institute, Wesleyan University, Middletown CT, June.

Shabazz, R. (2009) 'So high you can't get over it, so low you can't get under it': Carceral spatiality and black masculinities in the United States and South Africa. *Souls* II(3): 276–294.

Shabazz, R. (2015a) 'Sores in the city': A genealogy of the Almighty Black P. Stone Rangers. In Morin, K.M. and Moran, D. (eds) *Historical Geographies of Prisons: Unlocking the Usable Carceral Past*. London: Routledge, 51–67.

Shabazz, R. (2015b) *Spatializing Blackness: Architectures of Confinement and Black Masculinity in Chicago*. Urbana: University of Illinois Press.

Shubin, S. (1981) Research behind bars: Prisoners as experimental subjects. *Sciences* 21(1): 10–13.

Shukin, N. (2009) *Animal Capital: Rendering Life in Biopolitical Times*. Minneapolis: University of Minnesota Press.

Siebert, C. (2014) The rights of man … and beast. *New York Times Magazine*, 27 April.

Sinclair, U. (1906) *The Jungle*. New York: Doubleday.

Sittig, A. (2016) *The Mayans Among Us: Migrant Women and Meatpacking on the Great Plains*. Lincoln: University of Nebraska Press.

Skitolsky, A. (2008) Director, *La Historia de Nuestras Vidas*, Teatro Indocumentado, Decorah, Iowa.

Skloot, R. (2010) *The Immortal Life of Henrietta Lacks*. New York: Crown Publishers.

Smith, J. (2014) We are not the only political animals. *New York Times*, 2 November.

Smith, M. (2002) The ethical space of the abattoir: On the (in)human(e) slaughter of other animals. *Human Ecology Review* 9(2): 49–58.

Smith, M. (2017) Minnesota officer acquitted in killing of Philando Castile. *New York Times*, 16 June.

Socha, K. (2013) The 'dreaded comparisons' and speciesism: Leveling the hierarchy of suffering. In Socha, K. and Blum, S. (eds) *Confronting Animal Exploitation: Grassroots Essays on liberation and Veganism*. Jefferson NC: McFarland & Co., 223–239.

Socha, K. and Blum, S. (eds) (2013) *Confronting Animal Exploitation: Grassroots Essays on Liberation and Veganism*. Jefferson NC: McFarland & Co.

Solitary Watch: News from a Nation on Lockdown (2013) Voices from solitary: Death row diary of Florida man scheduled to die tonight. 12 June. Available at: http://solitary-watch.com/.

Sontag, S. (2003) *Regarding the Pain of Others*. New York: Farrar, Straus, and Giroux.

Sperry, R. (2014) Death by design: An execution chamber at San Quentin State Prison. *The Avery Review* 2: October. Available at: http://averyreview.com/issues/2/death-by-design.

Spiegel, M. (1997) *The Dreaded Comparison: Human and Animal Slavery*. New York: Mirror Books.

Srinivasan, K. (2013) The biopolitics of animal being and welfare: Dog control and care in the UK and India. *Transactions of the Institute of British Geographers* 38: 106–119.

Steiker, C.S. and Steiker, J.M. (2014) The death penalty and mass incarceration: Convergences and divergences. *American Journal of Criminal Law*, 1 April.

Steiner, G. (2005) *Anthropocentrism and Its Discontents: The Moral Status of Animals in the History of Western Philosophy*. Pittsburgh PA: University of Pittsburgh Press.

Steiner, G. (2008) *Animals and the Moral Community: Mental Life, Moral Status, and Kinship*. New York: Columbia University Press.

Steiner, G. (2013) *Animals and the Limits of Postmodernism*. New York: Columbia University Press.

Steiner, G. (2016) Personal correspondence with author, December.

Stepansky, J., Parascandola, R., Morales, M., and Otis, G. (2014) Helpless stray cat goes flying after being kicked by heartless creep in Brooklyn. *New York Daily News*, 6 May.

Stobbe, M. (2011) Ugly past of U.S. human experiments uncovered. Associated Press. Available at: www.nbcnews.com/id/41811750/ns/health-health_care/t/ugly-past-us-human-experiments-uncovered/#.VsTeIPIrKUk.

Striffler, S. (2005) *Chicken: The Dangerous Transformation of America's Favorite Food*. New Haven CT: Yale University Press.

Supreme Court upholds lethal injection procedure (2015) *Washington Post*, 29 June.

Tait, P. (2011) *Wild and Dangerous Performances: Animals, Emotion, Circus*. New York: Palgrave Macmillan.

Talvi, S.J.A. (2002) The prison as laboratory: Experimental medical research on inmates is on the rise. *In These Times*, 23 January.

Taylor, C. (2013) Foucault and critical animal studies: Genealogies of agricultural power. *Philosophy Compass* 8(6): 539–551.

Taylor, N. (2011) Criminology and human-animal violence research: The contribution and the challenge. *Critical Criminology* 19(3): 251–263.

Taylor, S. (2013) Vegans, freaks, and animals: Toward a new table fellowship. *American Quarterly* 65(3): 757–764.

Thomas, K. (1983) *Man and the Natural World: Changing Attitudes in England 1500–1800*. London: Allen Lane.

Thomas, S. and Shields, L. (eds) (2012) Prison studies and critical animal studies: Understanding interconnectedness beyond institutional comparisons. *Journal for Critical Animal Studies* 10: 4–11.

Thompson, H.A. (2016) *Blood in the Water: The Attica Prison Uprising of 1971 and Its Legacy*. New York: Pantheon.

Toobin, J. (2016) The strange case of the American death penalty. *New York Times*, 21 December.

Tuan, Y-F. (1984) *Dominance and Affection: The Making of Pets*. New Haven CT: Yale University Press.

Tuan, Y-F. (1999) Geography and evil: A sketch. In Proctor, J. and Smith, D. (eds) *Geography and Ethics: Journeys in a Moral Terrain*. London: Routledge, 106–119.

Turner, J. (2013) The politics of carceral spectacle: Televising prison life. In Moran, D., Gill, N., and Conlon, D. (eds) *Carceral Spaces: Mobility and Agency in Imprisonment and Migrant Detention*. London: Ashgate, 219–237.

Tyner, J.A. and Colucci, A.R. (2015) Bare life, dead labor, and capital(ist) punishment. *ACME: An International Journal for Critical Geography* 14(4): 1083–1099.

Uddin, L. (2015) *Zoo Renewal: White Flight and the Animal Ghetto*. Minneapolis: University of Minnesota Press.

Union of Concerned Scientists (2016) Hidden costs of industrial agriculture. Available at: www.ucsusa.org/food_and_agriculture/our-failing-food-system/industrial-agriculture/hidden-costs-of-industrial.html#.WGfTFvP-FTc.

Urbanik, J. (2012) *Placing Animals: An Introduction to the Geography of Human–Animal Relations*. Lanham MD: Rowman & Littlefield.

U.S. Department of Health and Human Services (2017) *Observations on Prescription Drug Spending*. Available at: https://aspe.hhs.gov/pdf-report/observations-trends-prescrip tion-drug-spending.

Vanyur, J. (1995) Design meets mission at new federal max facility. *Corrections Today* 57: 90–97.

Vickers, B. (2014) Three and a half steps. Available at: www.hankskinner.org/hs/hs.php?en,3-halfsteps.

Wacquant, L. (2001) 'Deadly symbiosis': When ghetto and prison meet and mesh. In Garland, D. (ed) *Mass Imprisonment: Social Causes and Relations*. London: Sage, 293–304.

Wacquant, L. (2005) The great penal leap backward: Incarceration in America from Nixon to Clinton. In Pratt, J., Brown, D., Brown, M., Hallsworth, S., and Morrison, W. (eds) *The New Punitiveness: Trends, Theories, Perspectives*. Cullompton, Devon: Willan Publishing, 3–26.

Wacquant, L. (2009) *Punishing the Poor: The Neoliberal Government of Social Insecurity*. Durham NC: Duke University Press.

Wadiwel, D. (2009) The war against animals: Domination, law and sovereignty. *Griffith Law Review* 18(2): 283–297.

Wadiwel, D. (2015) *The War Against Animals*. Boston: Brill Rodipi.

Washington, H. (2006) *Medical Apartheid: The Dark History of Medical Experimentation on Black Americans from Colonial Times to the Present*. New York: Doubleday.

Washington, H. (2012) *Deadly Monopolies: The Shocking Corporate Takeover of Life Itself – and the Consequences for Your Health and Our Medical Future*. New York: Anchor Books.

Watts, M.J. (2000) Afterword: Enclosure. In Philo, C. and Wilbert, C. (eds) *Animal Spaces, Beastly Places: New Geographies of Human–Animal Relations*. New York: Routledge, 293–304.

Waxman, S. (2016) Colin Dayan's ethics without reason. *Boston Review*, 12 January.

White, C. (2017) Black lives, sacred humanity, and the racialization of nature, or why America needs religious naturalism today. *American Journal of Theology & Philosophy*. 38 (2–3): 109–122.

Wilderson, F. (2010) *Red, White, and Black: Cinema and the Structure of U.S. Antagonisms*. Durham NC: Duke University Press.

Williams, T. (2016) U.S. Correctional population at lowest level in over a decade. *New York Times*, 29 December.

Wise, S. (2000) *Rattling the Cage: Toward Legal Rights for Animals*. New York: Basic Books.

Wise, S. (2002) *Unlocking the Cage*. New York: Basic Books.

Wise, S. (2004) Animal rights, one step at a time. In Sunstein, C.R. and Nussbaum, M.C. (eds) *Animal Rights: Current Debates and New Directions*. New York: Oxford University Press, 19–50.

Wise, S. (2009) *An American Trilogy: Death, Slavery, and Dominion on the Banks of the Cape Fear River*. Philadelphia PA: Da Capo Press, Perseus Books Group.

Wise, S. (2016) *Unlocking the Cage*. Film directed by Hegedus, C. and Pennebaker, D.A. First Run Films, Kanopy.

Wishart, D. (ed.) (2004) Cattle Towns and Frontier Violence. In Wishart, D., *Encyclopedia of the Great Plains*. Lincoln: University of Nebraska Press, 162, 387.

Witness to an Execution (2000) Sound Portraits.com. Available at: http://soundportraits. org/on-air/execution_tapes/.

Wolch, J. (1998) Zoopolis. In Wolch, J. and Emel, J. (eds) *Animal Geographies: Place, Politics and Identity in the Nature–Culture Borderlands*. New York: Verso, 119–139.

Wolch, J. and Emel, J. (eds) (1998a) *Animal Geographies: Place, Politics and Identity in the Nature–Culture Borderlands*. London: Verso.

Wolch, J. and Emel, J. (1998b) Borderlands. In Wolch, J. and Emel, J. (eds) *Animal Geographies: Place, Politics and Identity in the Nature–Culture Borderlands*. London: Verso.

Wolfe, C. (2003) *Animal Rites: American Culture, the Discourse of Species, and Posthumanist Theory*. Chicago: University of Chicago Press.

Wolfe, C. and Elmer, J. (1995) Subject to sacrifice: Ideology, psychoanalysis, and the discourse of species in Jonathan Demme's *Silence of the Lambs* . *Boundary 2* 22(3): 141–170.

Wolfers, J., Leonhardt, D., and Quealy, K. (2015) 1.5 Million missing black men. *New York Times*, 20 April.

Wolff, N., Blitz, C.L., Shi, J., Siegel, J., and Bachman, R. (2007) Physical violence inside prisons: Rates of victimization. *Criminal Justice and Behavior* 34: 588–599.

Wolfson, D.J. and Sullivan, M. (2004) Foxes in the hen house: Animals, agribusiness, and the law. A modern American fable. In Sunstein, C.R. and Nussbaum, M.C. (eds) *Animal Rights: Current Debates and New Directions*. New York: Oxford University Press, 205–228.

Womack, C. (2013) There is no respectful way to kill an animal. *Studies in American Indian Literature* 25(4): 11–27.

Woodward, K. and Lea, J. (2010) Geographies of affect. In Smith, S. J., Pain, R., Marston, S.A., and Jones III, J.P. (eds) *The SAGE Handbook of Social Geographies*. London: Sage, 154–175.

Wrye, J. (2015) 'Deep inside dogs know what they want': Animality, affect, and kill-ability in commercial pet foods. In Lopez, P. and Gillespie, K. (eds) *Economies of Death: Economic Logics of Killable Life and Grievable*. London: Routledge, 95–114.

Wynter, S. (2003) Unsettling the coloniality of being/power/truth/freedom: Toward the human, after man, its overrepresentation – an argument. *CR: The New Centennial Review* 3(3): 257–337.

Yildirum, M. (2017) Mapping the carceral geography of 'street dog reservoirs' in Istanbul. Paper presented at the 7th Annual Nordic Geographers Meeting, Stockholm, Sweden.

Yoram, C. (2002) Cruelty to animals' and nostalgic totality: Performance of a traveling circus in Britain. *The International Journal of Sociology and Social Policy* 22(11/12): 73–88.

Zellhoefer, A. (2013) Animal enterprise acts and the prosecution of the 'SHAC 7': An insider's perspective. In Socha, K. and Blum, S. (eds) *Confronting Animal Exploitation: Grassroots Essays on Liberation and Veganism*. Jefferson NC: McFarland & Co., 241–254.

Zimmermann, A., Hatchwell, M., Dickie, L., and West, C. (eds) (2007) *Zoos in the 21st Century: Catalysts for Conservation?* Cambridge: Cambridge University Press.

Zinn, H. (1999) *A People's History of the United States: 1492–Present*. New York: HarperCollins.

Index

Page numbers in *italics* show an illustration, n indicates an endnote

Yerkes Center, Atlanta 72
Yildirum, Mine 102

Zellhoefer, Aaron 72–3
zoos: alternative models 130–131, *131*,
142n2–143n3; animal rights activism
and reform 136–137, 138; cage and
prison cell parallels 120–121, 124, 139;

captivity, negative experiences
125–127, 139; conservation mission
failures 129–130; developmental
stages 128–129; gorilla death,
Cincinnati Zoo 15, 17, 140, 141;
projecting 'wild animal' links 122–123;
visibility of confinement 123–125,
128–129, 140

.

Milton Keynes UK
Ingram Content Group UK Ltd.
UKHW040053071024
449327UK00019B/543